Exploring Trigonometry

With

THE GEOMETER'S SKETCHPAD®

David Shaffer

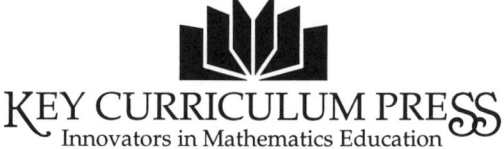

Exploring Trigonometry
with *The Geometer's Sketchpad*

Author: David Shaffer

Editing and Production: Rob Berkelman, William Finzer

Sketchpad Design and Implementation: Nicholas Jackiw

Project Direction: William Finzer

Dedication: For Amy,
  For ever

The Geometer's Sketchpad is a product of the Visual Geometry Project at Swarthmore College. Portions of the work on this book were funded by the National Science Foundation. The Visual Geometry Project was directed by Drs. Eugene Klotz and Doris Schattschneider.

This material is based upon the work supported by the National Science Foundation under award number ISI-9060238. Any opinions, findings, and conclusions or recommendations expressed in this publication are those of the author and do not necessarily reflect the views of the National Science Foundation.

The Geometer's Sketchpad and *Dynamic Geometry* are trademarks of Key Curriculum Press, Inc. All other brand names and product names are trademarks or registered trademarks of their respective holders.

© 1995 Key Curriculum Press, Inc. All rights reserved.
ISBN 1-55953-075-8

10 9 8 7 6 5 4 3   99 98 97 96

Key Curriculum Press
P.O. Box 2304
Berkeley, California 94702
510-548-2304

Exploring Trigonometry Sketches Disks

Key Curriculum Press guarantees that the Sketches and Scripts Disks that accompany this book are free of defects in materials and workmanship. A defective disk will be replaced free of charge if returned within 90 days of the purchase date. After 90 days, there is a $10.00 replacement fee.

Limited Reproduction Permission

Key Curriculum Press grants the teacher who purchases *Exploring Trigonometry with The Geometer's Sketchpad* the right to reproduce activities and example sketches and scripts for use with his or her own students. Unauthorized copying of *Exploring Trigonometry with The Geometer's Sketchpad* or of the *Exploring Trigonometry* sketches is a violation of federal law.

Contents

Section 1: Introduction 1
    The Origins of *Exploring Trigonometry* 1
    How This Book Is Organized 2
    Supplementary Sketches 2
    How Explorations Are Organized 3
    Resources Available for Teachers 4
    How to Use the Explorations 4
    General Suggestions for Using Explorations 7

Section 2: Basic Trigonometry 9
    Exploration 2.1: The Area of a Triangle 10
    Exploration 2.2: Function by Analogy 14
    Exploration 2.3: Special Angles 20
    Exploration 2.4: Raul's Ride 25

Section 3: Fundamental Functions 31
    Exploration 3.1: Area Revisited 32
    Exploration 3.2: How High Is It? 37
    Exploration 3.3: Reading a Graph 42
    Exploration 3.4: Air on the P String 46

Section 4: Laws and Their Applications 51
    Exploration 4.1: Navigation 52
    Exploration 4.2: Invariant Quantities 56
    Exploration 4.3: Special and General 61
    Exploration 4.4: Ambiguity 65

Section 5: Extended Investigation—Motion of a Pendulum 71
    Exploration 5.1: Building a Pendulum 72
    Exploration 5.2: Modeling a Pendulum 76
    Exploration 5.3: Gravity 80
    Exploration 5.4: Simple Harmonic Motion 84

Section 6: Introduction to Polar Coordinates 89
    Exploration 6.1: Cartesian Coordinates 90
    Exploration 6.2: Polar Coordinates 94
    Exploration 6.3: Trigonometry and Coordinate Systems 97

Section 1: Introduction

The Origins of *Exploring Trigonometry*

The explorations that make up this book have relatively humble origins. As I was teaching a semester-long trigonometry course for high school juniors and seniors, I was looking for a way to give them the kind of visual intuition about circles and angles that forms the underpinnings of my own understanding of trigonometry. I wanted them to picture a unit circle with a right triangle in it when they thought about the sine function. I wanted them to *see* in their mind's eye why cosine can never be greater than one.

The tool I chose for the task of creating visual intuition was The Geometer's Sketchpad. The process meant a number of trips during the semester to our school's rather ill-equipped computer lab. At first I used Sketchpad with our one overhead display computer, showing students some of the basic tools of the program such as how to make points and lines and how to measure angles. After a brief introduction I asked students to construct their own unit circle models based on my demonstration.

Mostly I wanted students to play with their model unit circles. My goal was to build their intuitive understanding of the relationships that form the basis of trigonometry. I wanted them to get a feel for how the *y*-coordinate of a point on a circle changes as the central angle changes. I wanted them to see why the sine of an angle is independent of the hypotenuse of a triangle.

Of course, they needed some guidance. To build their intuition they needed to observe, reflect, and conjecture about their experiments. They needed to play attentively. So I provided a series of questions for them to explore, then asked them to form and test hypotheses based on their answers. I asked them to generalize from their specific insights. Many of the early explorations in this collection are the direct descendants of these early attempts.

This process of structured play and leading questions is one that I came to call *guided discovery*. Students are not "discovering" the Law of Sines when they construct a triangle, measure sides, calculate sines, and compute ratios. Yet it feels like a discovery to them when they realize that from their own calculations comes a constant ratio, a rule or a trick that they have seen for themselves that they can use to solve problems. It is not discovery in the strictest sense, but it provides students with a sense of ownership and a taste for deeper investigation.

My goal was to provide a rich set of resources to complement a traditional trigonometry curriculum. Just as I used activities to give students an intuitive understanding of circles, angles, and triangles, the explorations are designed to give students insight into the key concepts of trigonometry.

Of course, teachers face differing constraints in their classrooms. Some have access to adequate resources for using computer software; others do not. Some have large classes; some have small groups of students. Some have bright, mathematically gifted students; some have students for whom mathematics is a particularly difficult challenge. Because of this range, the explorations are designed to support a variety of teaching and learning styles. They can be used as notes for a class lecture with demonstrations. They can be used for group investigation by students. They can be used as a springboard for open-ended discussion and discovery. Some teachers may choose to introduce new concepts using the explorations and follow up with more traditional practice exercises. Others may use the explorations to provide a deeper look at ideas already discussed in class.

However one chooses to use the activities in *Exploring Trigonometry*, the limitation of these explorations is that they do not constitute a textbook. They are not designed to stand alone as a course in trigonometry. Rather, they are intended to supplement a coherent program of instruction, providing at key points a deeper understanding of the basic processes in the theory and practice of trigonometry.

How This Book Is Organized

Exploring Trigonometry is organized into five sections. Each section contains three or four explorations in a single area of trigonometry. A brief introduction to the explorations is at the beginning of each section. The introduction describes the content of the explorations and the way they relate to one another and to other sections of the book.

In many cases, the explorations can be done in any order according to the organization of your textbook or course syllabus. In some cases, such as the extended investigation in Section 5, explorations are designed to be done in sequence. When this is the case, the section introduction makes specific suggestions about the order and timing of the explorations.

In general, the explorations at the beginning of the book assume less familiarity with Sketchpad than the later explorations. When an exploration requires specific skills or mathematical knowledge, the requirements are included in the introduction to each section.

Supplementary Sketches

Included on the accompanying disk are many additional sketches and scripts which can be useful to you in your class—either as demonstration sketches or as ideas for student projects. The disk also includes a Read Me file containing descriptions of these sketches and scripts. These sketches and scripts include a working model of the unit circle, a directed distance script, a visual proof of the laws of sine and cosine, a working model of the sine wave, and some sketches and scripts for working with vectors.

How Explorations Are Organized

The explorations in this book follow a common sequence, modeled very loosely on the steps of problem solving outlined in George Polya's classic, *How to Solve It*. Students create a sketch, investigate the relationships it models, generalize from their observations, answer specific questions, reflect on their understanding, and explore other related questions.

Each exploration begins with the construction of a sketch. In the Construct section, students are given an explicit set of steps for creating a model they will examine in the exploration. The process of construction is often extremely valuable for building understanding about the mathematics of a situation; it takes students time to construct their own models, but it is time well spent.

In the Investigate section, students are asked to change or manipulate their model in various ways that help them see the underlying relationships the model represents. They often answer specific questions, some of which require them to grapple with new concepts or figure out relationships they have not seen before.

After their investigation, students are asked to form a conjecture based on their observations. Conjectures may be speculations based on initial observations. They may be questions for further examinations. They may be insights that were prompted by the investigation of the model. Conjecturing about mathematical observations is an expansive process, in which students are asked to think broadly and inductively about the material they are exploring.

The Construct, Investigate, and Conjecture sections are the basis of each exploration. You may want to first hand these sections out to students, then discuss their conjectures individually or in groups before proceeding to the next sections. This will encourage students, especially early on, to make their own conjectures about their observations.

Following the first three sections of each exploration, there is a page of follow-up questions. These help students connect their observations and insights to the formal structures of trigonometry and to other aspects of the study of mathematics.

The Focus section asks specific questions that help students connect their investigation to the standard equations and functions of trigonometry. So, for example, students write their own formulation of the Law of Cosines based on their investigation or define the tangent function in terms of the sides of a triangle.

The Look Back section provides students with an opportunity to reflect on their investigation and the specific results of the Focus section. It asks students to tie their discoveries to other concepts in trigonometry, mathematics generally, or the world around them.

 The Explore More section provides additional open-ended questions that ask students to continue thinking about their discoveries in new ways or in broader contexts. The Explore More questions are useful for more extended projects, or for extra-credit work for advanced students.

Resources Available for Teachers

Following each exploration is a page of teacher's notes. The notes follow the section organization of the explorations. The notes include suggestions and advice on the use of Sketchpad, points where students may encounter particular difficulties, suggested responses to common student questions, sample answers, topics for discussion, and other material that may be useful when working with students on a specific exploration.

Included with this book are two disks of sample sketches. Some of the explorations require premade sketches (see, for example, Exploration 2.4: Raul's Ride). These sketches are included on the disks, along with some completed sketches based on the Construct section of the explorations, and some additional sketches that contain sample demonstrations on aspects of trigonometry not directly covered in the explorations in this book.

Before assigning an exploration to students, you may want to work through the various sections using the teacher's notes and sample sketches. This will help you anticipate questions from students that may not be covered explicitly in the supporting materials.

How to Use the Explorations

The explorations in this book can be used in a number of different ways, depending on the time and facilities available and the goals of the course being taught. In general, teachers need to decide (1) how students should conduct a given exploration, and (2) what students will produce at the end of the exploration. Depending on the level of your students and your objectives for individual activities, you may want to try different teaching and assessment strategies for different explorations within the same course.

How to Conduct an Exploration

There are a number of ways to work through the explorations. Some examples are listed below, along with an indication of the amount of class time you can expect to spend on one exploration using each method. Of course, times will vary depending on the specific situation in your classroom.

1. *Individual Investigation*
   Students work on their own through all steps of the exploration, using their own computer and consulting with the teacher where necessary. This method can also be used for assessment if students are required to turn in written answers to substantial portions of the exploration.
   Advantages: Students master material on their own and are free to work at their own pace.
   Disadvantages: Requires more time and equipment than other methods. May be intimidating for weaker students.
   Expected Time: 3–4 class periods, or 120–200 minutes.

2. *Group Investigation*
   Students work in groups of two to four through all steps of the exploration, using a single computer and consulting with one another and with the teacher.
   Advantages: Students benefit from insights of peers. Students learn to work in group situations.
   Disadvantages: Requires more time than other methods. Weaker students may not participate fully in large groups.
   Expected Time: 2 class periods, or 80–100 minutes.

3. *Investigation with Class Discussion*
   Students work on Construct, Investigate, and Conjecture sections individually or in groups. Then the teacher leads class discussion, including discussion of student conjectures, Focus, and Look Back questions.
   Advantages: Teacher can address student conjectures. Students get uniform answers to Focus and Look Back questions.
   Disadvantages: Students may have less of a sense of ownership of the material. May require use of demonstration equipment.
   Expected Time: 1–2 class periods, or 40–80 minutes.

4. *Demonstration with Follow-Up*
   Teacher uses demonstration computer to lead class through Construct, Investigate, and Conjecture sections. Students answer Focus and Look Back questions individually or in groups.
   Advantages: Requires less class time than other methods. Teacher can model good investigation technique.
   Disadvantages: Requires demonstration equipment; reduces students' sense of discovery.
   Expected Time: 1–2 class periods, or 40–80 minutes.

5. *Lecture with Demonstration*
   Teacher leads class through Construct and Investigate sections, followed by a class discussion of Conjecture, Focus, and Look Back sections.

Exploring Trigonometry ©1995 by Key Curriculum Press

Advantages: Requires less time than other methods. Students get uniform answers to Focus and Look Back questions.
Disadvantages: Requires demonstration equipment. Students do not use Sketchpad on their own.
Expected Time: 1 class period, or 40–50 minutes.

What Students Produce in an Exploration

There are a number of ways to assess student work in explorations. Some examples are listed below. You may want to use a combination of these or other methods.

1. *Direct Observation*
   Teacher observes students as they work on the exploration. Assessment is based on specific criteria announced to students beforehand, such as ability to work productively with others, ability to record observations accurately, and ability to explain and discuss ideas and insights.
   Advantages: Teacher can assess process as well as product. Students do not need to do additional work writing up their results.
   Disadvantages: Teacher cannot observe all work of all students. Teacher is not available to help students while they work.

2. *Project-Based Assessment*
   Students complete the exploration and then produce a project based on the material covered in the exploration. Projects may be posters, reports, research, invented problems, or real-world applications.
   Advantages: Connects mathematical discovery to other aspects of life and learning.
   Disadvantages: May not cover all aspects of the content of the exploration.

3. *Indirect Assessment of Content*
   Teacher asks students to complete exploration, after which students take a test or demonstrate mastery of specific types of problems based on the material of the exploration.
   Advantages: Students are responsible for mastering the material of the exploration.
   Disadvantages: Focuses on product only and not process.

4. *Answers to Key Questions*
   Selected questions from the exploration are discussed through student presentation or through teacher interview of students. The questions may or may not be announced beforehand.
   Advantages: Focuses on students' ability to discuss mathematics. Makes students accountable for key ideas.

Disadvantages: Requires additional class time. May be intimidating to some students.

5. *Lab Report*
   Students write up a full report of their exploration, including data collected, observations made, and answers to all questions.
   Advantages: Provides a comprehensive picture of student work on an exploration. Encourages students to give specific answers to exploration questions.
   Disadvantages: Requires considerable time outside of class for teacher and students.

General Suggestions for Using Explorations

All of the above is meant to help you, the teacher, find a method to help students start discovering mathematics using the explorations in this book. There is clearly no one right way to go about that difficult and rewarding task. Trite but true: Whatever works for you is right for you.

In the hopes that at least one specific model will be helpful as you think about how to use the explorations in this book, I have tried to describe below some of my own thoughts about *Exploring Trigonometry* based on the way I use it in my own classes. Most of what I have learned about using guided discovery I learned by trial and error, and you may find that it takes one or two explorations before you have a sense of what works for your students.

I have used most of the methods of conducting and evaluating explorations described above in some form or another. I tend to have students work in groups of two or three because I find that they benefit from having someone with whom they can discuss difficult steps in the investigation or complex ideas. Also, I do not usually have enough computers to have students work alone. Most of the time I let students choose their own working groups.

I tend to use explorations to introduce topics, discussing the mathematics *after* students have tried to discover on their own. On occasion, I use explorations as evaluation, asking students to show their mastery of a concept by successfully completing an exploration.

I find that I do about one third of the explorations as demonstrations rather than as independent investigations, usually because of the pressure of time. Students seem to do well with about one independent investigation every one or two weeks, though I tend to do the investigations in Section 5 (the pendulum lab) and Section 6 (polar coordinates) in groups as extended one- or two-week projects.

I often ask students to write up an extensive lab report on their explorations (especially for week-long projects), or I give them the option of turning in a report for extra credit. I almost always follow explorations up with practice problems and a quiz or test. Students sometimes have a tendency to gloss over answers to difficult questions. The requirement that they write or explain their answers in a formal setting encourages them to dig deeper. I try to make them accountable for the content and connections of the investigation in an essay, a problem on a test, or written answers to specific questions.

Finally, I find that students enjoy the process of discovery, but that they also need help connecting their observations to their broader body of mathematical knowledge. I usually try to follow explorations with time for discussion of more general questions and issues. When students have solid intuitive and practical understanding of a specific concept, they are often eager to see how their new knowledge fits into a bigger picture. Some of my most enjoyable and successful lectures have concluded classes in which students discovered concepts on their own as a result of completing the activities in *Exploring Trigonometry*.

I hope you find these explorations as helpful to your students and as invigorating to your classes as I have to mine.

David Shaffer
Castilleja School
Palo Alto, California

Section 2: Basic Trigonometry

There are four explorations in this section of *Exploring Trigonometry*. Together they introduce students to the basic concepts of trigonometry: circles, triangles, angles measured in degrees and radians, and special angles.

Exploration 2.1: The Area of a Triangle is an introductory exploration, designed for students unfamiliar with Sketchpad. It also reviews some basic geometric ideas used in trigonometry, including base, height, and area of a triangle. This exploration contains no unfamiliar material, and if students are proficient in the use of Sketchpad, you might consider beginning with Exploration 2.2.

Exploration 2.2: Function by Analogy builds on the review of basic concepts in Exploration 2.1. Students use Sketchpad to produce the unit circle definition of the sine function. If you plan to use other explorations in this section, Exploration 2.2 should probably follow Exploration 2.1 and precede Exploration 2.3. Exploration 2.2 works well for students who have never seen the sine function before. It also helps students see how the triangle definition of sine (sine is "opposite over hypotenuse") is equivalent to the unit circle definition.

Exploration 2.3: Special Angles assumes that students understand the definition of the sine function. In Exploration 2.3 students use Sketchpad to discover a pattern in the values of the sine function at the special angles of 0°, 30°, 45°, 60°, and 90°. Exploration 2.3 works well as a discovery activity done in small groups or as an assessment activity if students are familiar with Sketchpad.

Exploration 2.4: Raul's Ride requires basic proficiency with Sketchpad but does not depend on the other explorations in Section 2. Students use Sketchpad to measure angles in radians and to convert between degree and radian measure. Exploration 2.4 can be used as a basic introduction to radian measure or as a challenge activity for students who are already familiar with radian and degree measure.

Try the sketch Unit Circle (*Mac*) or UnitCrc (*Windows*)—it's one of the many fun and useful supplementary sketches on the accompanying disk.

Exploration 2.1: The Area of a Triangle

Circles, triangles, and angles are mathematical concepts that are probably familiar to you from your study of algebra and geometry. Trigonometry describes the ways in which circles, triangles, and angles are related. The word *trigonometry* comes from the Greek words *trigonon*, meaning "triangle," and *metron*, meaning "measure."

In this exploration you will look at one important way to measure a triangle: by finding its *area*. You will also discover one way that the area of a triangle can be found from a circle.

Construct

For this exploration you will need to construct a circle at the origin of two perpendicular axes, a line parallel to the horizontal axis and tangent to the circle, and a triangle between the two parallel lines.

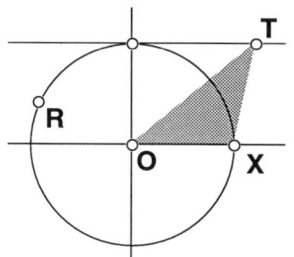

You may be tempted to construct this sketch from the picture without following the steps below. Be careful: Some of the objects in your sketch may not move as they are supposed to if you skip the instructions!

C1. Draw a horizontal line and a vertical line in the center of your sketch. Hide the points on these lines (the *control points*).

C2. Construct a medium-sized circle centered at the intersection of the two lines. Label the center point *O* and the point on the circle (the radius point) *R*.

C3. Construct point *X* at the right-side intersection of the circle and the horizontal line.

C4. Construct a point at the top intersection of the circle and the vertical line.

C5. Construct a line through this point parallel to the horizontal line. To do this, select the point and the horizontal line and choose Parallel Line from the Construct menu.

C6. Construct point *T* on the parallel line.

C7. Construct triangle *OXT* by selecting points *O*, *X*, and *T*. Then choose Polygon Interior from the Construct menu.

Investigate

I1. Measure the radius of the circle. To do this, select the circle and choose Radius from the Measure menu.

I2. Measure the area of the triangle. To do this, select the triangle interior and choose Area from the Measure menu.

I3. Move point *T*. Describe how the area of the triangle and the radius of the circle change as you move point *T*.

I4. Move point *R*. Describe how the area of the triangle and the radius of the circle change as you move point *R*. Explain why this happens.

I5. Describe in general terms the relationship between the area of the triangle and the radius of the circle.

To measure the distance between two points, select both points and choose Distance from the Measure menu.

I6. Measure the length of the base of the triangle.

I7. Measure the height of the triangle.

I8. Calculate the area of the triangle from the base and the height. To do this, select the *measurement* of the base and the *measurement* of the height, then choose Calculate from the Measure menu. Use Sketchpad's calculator to calculate the area using the measurements on the calculator's pull-down menu.

I9. Write an equation for the area of the triangle in terms of the radius of the circle.

 Conjecture

Record any insights that you had during your investigation. What things that you believe are true in general are based on your specific observations? What patterns did you notice that you cannot explain yet? What questions would you like to explore further?

Follow-up—Exploration 2.1: The Area of a Triangle

 Focus

F1. Use Sketchpad's calculator to calculate the value of the expression shown below.

$$\frac{\text{area of the triangle}}{(\text{radius})^2}$$

How does this number change as you move points T and R? Where does this number come from?

F2. Use your observations to *prove* that the equation you wrote in step I9 for the area of the triangle in terms of the radius is correct.

 Look Back

L1. Why is it important that the line containing point T is parallel to the first horizontal line in the sketch? How would the sketch and the equation you wrote in I9 change if the line containing point T were *not* parallel to the horizontal line?

L2. Does it matter that the horizontal and vertical lines you drew at the beginning of your construction are perpendicular? How would the sketch and the equation you wrote in I9 change if they were not perpendicular?

 Explore More

E1. Construct a parallelogram with points T, O, and X as three of the vertices. What would be the relationship between the area of the parallelogram and the radius of the circle?

Teacher's Notes—Exploration 2.1: The Area of a Triangle

Construct

Students who have never done explorations using Sketchpad often try to draw the sketch from the diagram alone. The sketch they produce may look correct, but because the objects are not defined in the correct relationships, the sketch does not move as it is supposed to in the Investigate section. You may want to suggest that students follow the directions carefully until they are more familiar with Sketchpad and the explorations.

C5. Students sometimes need help selecting more than one object. Show them how to click in the empty space on the sketch to clear their selections. You may need to show them how to hold the Shift key while they click on more than one object.

Investigate

I3. Students are sometimes confused by the fact that neither the area nor the radius change when they move point T. Some students may feel that the question is asking them to find something that changes and assume that they have done something wrong. You may need to encourage them to have confidence in their results. Try suggesting that neither the radius nor the area change is an appropriate description of how these quantities change.

I8. Encourage students to move point T and point R after they have computed the area. They should observe how Sketchpad automatically recalculates the area for them based on the changing length of the base and height.

Conjecture

If this is students' first exploration, you may want to refer to the description of the conjecture question in *Section 1: Introduction* for hints on how to help them produce good conjectures. One good conjecture for this investigation might be that the point T does not affect the area because point T is on a line parallel to the base of the triangle; thus the height of the triangle is constant and so is the area.

Focus

F2. You will need to decide what you will accept as *proof* in this case. Students may simply make a table of values from Sketchpad, providing evidence of the validity of their equation. You may want to encourage them to write up a formal proof.

Look Back

L1. This question is particularly important in helping students to see why the sketch functions as it does.

Exploration 2.2: Function by Analogy

One of the most important ways our knowledge of mathematics grows is by *analogies.* A mathematical analogy is when two things that *appear different* are *similar* in some way—that is, they share something in common. By looking for more ways that things are the same, we often discover things we did not see when we looked at them separately.

For example, the way we multiply whole numbers is different from but very similar to the way we multiply fractions,

$$5 \times 6 = 30 \quad \text{and} \quad \frac{2}{3} \times \frac{7}{11} = \frac{2 \times 7}{3 \times 11} = \frac{14}{33}$$

We might wrongly expect by analogy that

$$\frac{2}{3} + \frac{7}{11} = \frac{2 + 7}{3 + 11} = \frac{9}{14}$$

Puzzling out why this analogy doesn't work can teach us a lot about fractions.

In this exploration, you will investigate another way to relate the area of a triangle to the size of a circle. Along the way, you will discover the most fundamental relationship in all of trigonometry.

Construct

For this exploration you will need to construct a circle at the origin of two perpendicular axes, a point on the circle, and a triangle inside the circle.

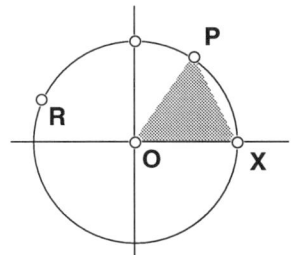

It is easier to draw vertical and horizontal lines if you hold down the shift key while drawing.

C1. Draw a horizontal line and a vertical line in the center of your sketch. Hide the points on these lines (the *control points*).

C2. Construct a medium-sized circle centered at the intersection of the two lines. Label the center point *O* and the point on the circle *R*.

If you are going to be doing the next exploration, you should save a copy of your sketch at this point to use then.

C3. Construct point *X* at the right-side intersection of the circle and the horizontal line.

C4. Construct point *P* on the circle.

14 • Section 2 ©1995 by Key Curriculum Press Basic Trigonometry

C5. Select points *P*, *O*, and *X*. Then choose Polygon Interior from the Construct menu.

Investigate

I1. Measure the radius of the circle and the area of the triangle.

I2. Change the radius of the circle. Describe the relationship between the area of the triangle and the radius of the circle.

It will make your sketch easier to understand if you relabel your measurements. To do this, choose the Text tool and double-click on your measurements. Use the dialog box to make the names shorter.

I3. Measure the **central angle** of the circle. The central angle is the angle of the triangle with vertex point *O*. To measure the central angle, select point *X*, point *O*, and point *P* in that order. Then choose Angle from the Measure menu.

I4. Move point *P*. Describe the relationship between the area of the triangle and the central angle of the circle. The relationship can appear complicated. Try describing it in general, *qualitative* terms.

To construct an altitude at point P that will remain perpendicular to the base of the triangle as you move point P, select the horizontal line and point P. Then choose Perpendicular Line from the Construct menu.

I5. Construct the altitude of the triangle at point *P*. Label the altitude segment *S-Height*.

I6. Measure the length of the base (the distance between the base vertices of the triangle) and the height (the length of the altitude) of the triangle.

I7. Calculate the area of the triangle from the base and the height. To do this, select the *measurement* of the base and the *measurement* of the height, then choose Calculate from the Measure menu. Use Sketchpad's calculator to calculate the area using the measurements on the calculator's pull-down menu.

To hide a measurement, select it, then choose Hide Measure from the Display menu.

I8. Check that the area you calculate is the same as the area Sketchpad measured for you. Then hide Sketchpad's measurement and use the value you calculated for the area of the triangle.

Exploring Trigonometry ©1995 by Key Curriculum Press

I9. Use Sketchpad's calculator to calculate the value of the expression below.

$$\frac{2 \times \text{area of the triangle}}{(\text{radius of the circle})^2}$$

I10. Change the radius of the circle. What do you notice?

I11. Move point *P* around the circle as you observe the expression you calculated in I9. Does it have a maximum and minimum value? Where is its value the greatest? Where is its value the least?

I12. Compare the length of segment *S-Height* to the value of the expression from I9 as point *P* moves around the circle. Can you find a relationship between the computed value and the length of segment *S-Height*?

I13. Use Sketchpad's calculator to calculate the **sine** of the central angle. The sine of an angle is the most basic function in trigonometry. To calculate the sine of the angle, select the *measurement* of the angle and then choose Calculate from the Measure menu. Choose Sin[from the calculator's pull-down menu. Then choose the measure of the angle from the pull-down menu. Then click on the) key on the calculator.

I14. Move point *P* around the circle. Record your observations.

Conjecture

Record any insights you had during your investigation. What things that you believe are true in general are based on your specific observations? What patterns did you notice that you cannot explain yet? What questions would you like to explore further?

Follow-up—Exploration 2.2: Function by Analogy

 Focus

F1. Use your observations to write an equation for the sine function of a central angle A in terms of the radius of the circle and segment S-Height.

F2. When reduced to simplest terms, the equation for the sine function in terms of the radius and segment S-Height is one way of defining the sine of an angle. Write up a precise explanation of how to find the sine of an angle using a circle for angles between 0° and 90°.

 Look Back

L1. Another way to define the sine of an angle is to construct a right triangle containing the angle. The sine is the length of the side opposite the angle divided by the length of the hypotenuse of the triangle. Show that your definition using a circle gives the same value as this right-triangle definition.

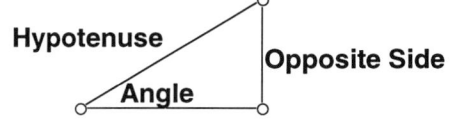

L2. How does the sketch and your definition of the sine function change if you move point R so that the radius of the circle is 1?

Explore More

E1. The relationship between the central angle and the area of the triangle is difficult to describe. One way mathematicians try to understand complicated relationships like this is to make a graph. What do you think a graph of the area of the triangle as a function of the angle would look like? Gather data from Sketchpad and make a graph.

After you make your graph, you might want to look at the sketch Sine Tracer (Mac) or Tracer (Windows).

E2. Could you make a graph of the length of segment S-Height as a function of the central angle? Gather data from Sketchpad and make a graph. How is the graph of segment S-Height different from the graph of the area of the triangle? How would the graph of segment *S-Height* compare to the graph of the *sine* of the central angle as a function of the central angle?

Teacher's Notes—Exploration 2.2: Function by Analogy

Construct

This sketch is used again in Exploration 2.3. You may want to encourage students to save a copy of their sketch before they begin the Investigation section.

C2. Students usually construct the circle using the Circle tool. This is fine as long as they label point R correctly. If they simply construct point R on the circle after the circle is constructed, point R will not control the radius of the circle as it should for the Investigate section.

Investigate

I3. Students who have not learned about angles in standard position may need some clarification about where the central angle is in this sketch.

I4. Students sometimes have trouble with qualitative descriptions. Encourage them to explain what they see. An example might be: As the central angle gets bigger, the area gets bigger until the central angle reaches 90°; as the central angle goes from 90° to 180°, the area gets smaller. A more sophisticated description might be: The area varies inversely with the difference between the central angle and 90°.

I5. Students sometimes draw the height in using a line segment. Make sure their height works no matter where they move point P on the circle. You can construct a segment for the height by constructing point H at the intersection of the x-axis and the perpendicular line through point P. Select point P and point H and construct a segment, then hide the perpendicular line.

I13. Students may notice that Sketchpad calculates positive values for the value calculated in I9. These values are equal to the absolute value of the sine function calculated in I13. This step provides an excellent opportunity to discuss the fact that the sine function is negative for angles between 180° and 360°. One explanation for why the sine is negative for some angles comes from the fact that when computing the sine function, the "height" of segment *S-Height* is usually considered to be the *directed* distance from line *x-axis* to point *P*—the distance is positive when point *P* is above the line and negative when point *P* is below the line. In other words, the directed length of segment *S-Height* is the *y*-coordinate of point *P*. You might want to point out to students that the value calculated in I9 is positive because Sketchpad cannot measure directed distance easily.

Conjecture

Note that the computed value is equal to the sine of the central angle. Also, the length of segment *S-Height* is equal to the radius multiplied by the sine of the central angle.

Focus

F2. Students may refer to the triangle definition of the sine function. You may want to encourage them to write their description without referring to *Opposite* and *Hypotenuse*. Show them how these terms become unclear for angles greater than 90°.

Look Back

The questions in this section are particularly good for making connections to other aspects of the study of trigonometry. Question L1 connects the triangle definition of the sine with the circle definition. Question L2 connects to the idea of a unit circle.

Explore More

These questions provide an excellent transition to the properties of the graphs of trigonometric functions. See also Explorations 3.1 and 3.3.

Exploration 2.3: Special Angles

Mathematicians use a particular notation—or shorthand—for the sine of an angle. Instead of writing the sine of an angle of 37 degrees, we usually write sin 37°. Sin 37° is read sine of thirty-seven degrees.

Not many years ago, students of trigonometry were taught to use sine tables—printed lists of values of the sine function—to determine the sine of angles. Because the tables were inconvenient and inexact, it was useful to learn and remember the value of the sine function at certain special angles. Now scientific calculators are so common and easy to use that there is not much need to memorize values. But there are some angles that come up often in the study of trigonometry (especially on tests!), and it is helpful to know the sine of a few angles so that you can make rough estimates and decide whether answers from your calculator are reasonable.

In this exploration, you will look at some special angles and discover a pattern that will help you remember their sine.

Construct

For this exploration you will need to construct a circle at the origin of two perpendicular axes, a point on the circle, and a triangle inside the circle.

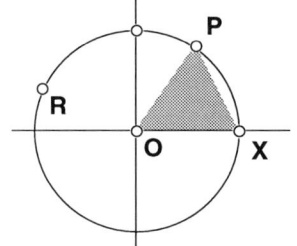

This construction is the same as the construction for Exploration 2.2. If you saved a copy of your construction, you can save time by using it.

C1. Construct a horizontal line and a vertical line.

C2. Construct a medium-sized circle centered at the intersection of the two lines. Label the center point O and the point on the circle R.

C3. Construct point X at the right-side intersection of the circle and the horizontal line.

C4. Construct point P on the circle.

C5. Select points P, O, and X, then choose Polygon Interior from the Construct menu.

Investigate

You may find it convenient to use a unit circle.

I1. Measure the radius of your circle, the height of the triangle at point P, and the central angle at point O.

I2. Use Sketchpad's calculator to calculate the sine of the central angle using the height of the triangle at point P and the radius of the circle.

You may want to use the Tabulate command on the Measure menu to help you.

I3. Move point *P* on the circle to determine values for sin 0°, sin 30°, sin 45°, sin 60°, and sin 90°. Describe any patterns you notice.

I4. Use Sketchpad's calculator to calculate (Sin[central angle])². Move point *P* to make the central angle 0°, 30°, 45°, 60°, and 90°. Describe any patterns you notice in the values of (Sin[central angle])².

I5. Calculate 4 × (Sin[central angle])². Move point *P* to make the central angle 0°, 30°, 45°, 60°, and 90°. Describe any patterns you notice.

 Conjecture

Record any insights you had during your investigation. What things that you believe are true in general are based on your specific observations? What patterns did you notice that you cannot explain yet? What questions would you like to explore further?

Follow-up—Exploration 2.3: Special Angles

 Focus

F1. Complete the chart below.

Central Angle A	Sin A	Sin A
0°	0	$\frac{\sqrt{}}{2}$
30°		$\frac{\sqrt{1}}{2}$
45°	.707	$\frac{\sqrt{}}{2}$
60°		$\frac{\sqrt{}}{2}$
90°		$\frac{\sqrt{4}}{2}$

F2. Use your observations to come up with a trick to help you remember the sine of 0°, 30°, 45°, 60°, and 90°. Describe your trick so that someone who had not done this exploration could remember it.

F3. Why is there no angle whose sine is $\sqrt{5}/2$?

 Look Back

You can construct these triangles using Sketchpad and measure angles and sides to get some insight.

L1. Use your knowledge of the Pythagorean Theorem, 30°-60°-90° triangles, and 45°-45°-90° triangles from geometry to prove that the values you found with Sketchpad are correct.

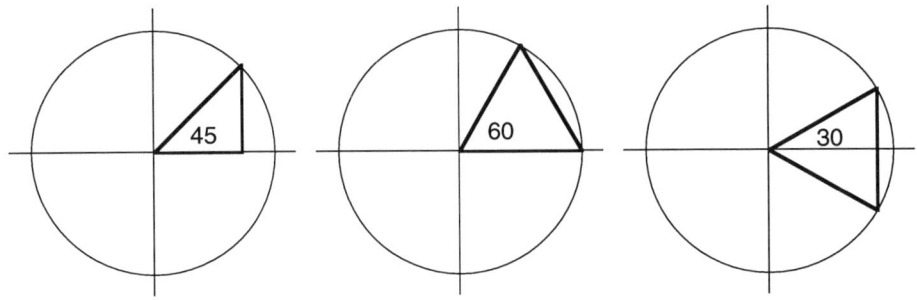

L2. Can you extend your trick (and your proof) to central angles of 120°, 135°, 150°, and 180°?

 Explore More

To get directed degrees, choose Preferences from the Display menu, then select directed degrees for degree measure in the dialog box.

E1. Have Sketchpad measure angles in directed degrees. Look at the values of sin (–30°), sin (–45°), and sin (–60°). What do you notice? Write your conjecture as an equation? Prove your conjecture is correct.

Teacher's Notes—Exploration 2.3: Special Angles

Construct

If students are working with a saved sketch, they can proceed directly to the Investigate section.

Investigate

Some students use their own calculators to compute values in the Investigate section. You may want to suggest to them that it is easier to use Sketchpad's calculator when multiple computations are involved because they will be able to see Sketchpad continuously update the calculations. Students may find some of the following questions easier to explore using Sketchpad's Tabulate command. The Tabulate command allows you to record and compare measurements as a sketch changes. To use the Tabulate command, select two or more measures to be compared; then choose Tabulate from the Calculate menu. Sketchpad will create a table showing the current values of the measures. Adjust the sketch so that the measures in the table show the values you want to record. Double-click on the table to save the values. You can then move the sketch again and enter additional values in the table by double-clicking.

I3. Some students become confused because they cannot find a simple pattern in the values of the sine function. You may need to reassure them that no simple pattern exists.

I4 Sketchpad may show sin 45° as .51 rather than .5. This provides an excellent opportunity to discuss the accuracy of measurements and rounding errors with students.

Conjecture

Students may see patterns in steps I4 and I5 but may not know what to make of them.

Focus

F1. If students are having trouble with the table, have them rewrite the equation in step I5 so that sine is alone on one side of the equation.

Look Back

L2. This question provides an opportunity to discuss the idea of reference angles.

Explore More

E1. One possible conjecture is that $\sin(x) = -\sin(-x)$.

Exploration 2.4: Raul's Ride

Raul had to take his kid brother Ernesto to the amusement park. He hated going with Ernesto because Ernesto was too scared to go on any of the fun rides. All Ernesto wanted to do was go on the carousel, over and over again. As they rode their carousel horses, Raul, being a clever student of trigonometry, decided to entertain himself by thinking about the circles and angles of the carousel as it turned. He imagined he rode his horse in a series of circles wrapped around and around the carousel. Raul wondered how far he had gone if all of these circles were stretched into a straight line. When he went to school the next day, he used Sketchpad to help him understand the carousel better.

In this exploration, you will use Raul's sketch to investigate the relationship between distance and angle and discover a new way to measure angles.

Construct

    Open the sketch Raul's Ride (Mac) or RaulRide (Windows).

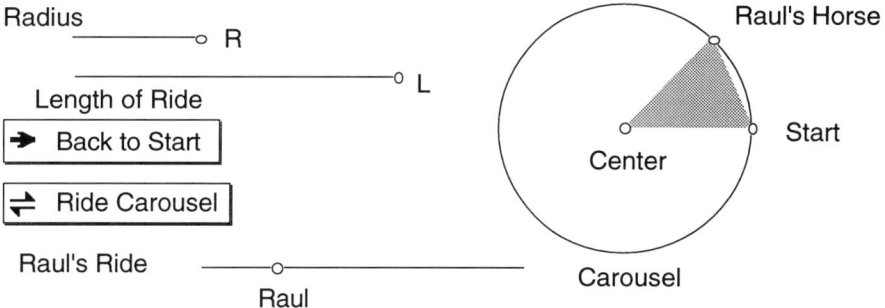

  C1. This sketch is set up to simulate Raul's ride around the carousel. To see how the sketch works, double-click on the Back to Start button. This moves the carousel back to its starting point and Raul back to the beginning of his distance counter.

  C2. Double-click the Ride Carousel button to start the carousel turning and to start Raul's distance counter.

Investigate

  I1. Watch what happens when you animate the sketch by double-clicking on the Ride Carousel button. Try changing the radius of the carousel or the length of Raul's ride. Double-click the Back to Start button.

Then double-click the Ride Carousel button. What changes do you notice in Raul's ride?

You may find it helpful to set the radius of the carousel to 1 cm.

I2. How long does Raul's ride have to be for him to make one complete trip around the carousel? How does the radius of the carousel affect your answer?

I3. Through what angle does Raul ride if he travels one fourth of the way around the carousel? Test your answer by measuring the central angle of the carousel. How does the radius of the carousel affect your answer?

I4. Through what *angle* does Raul ride if he travels a distance equal to the radius of the circle? How does the radius of the carousel affect your answer?

> **Definition**: One **radian** is the measure of an angle formed by an arc whose length is equal to the radius of a circle.

You can check your work by choosing Preferences from the Display menu; then use the dialog box to set degree measurements to radians. If you do this, be sure to adjust the precision for measuring angles to tenths or hundredths.

I5. How many degrees are there in an angle of *two* radians?

I6. How many degrees are there in an angle of π radians?

I7. Set the radius of the carousel to 1 cm. To have Sketchpad measure in centimeters, choose Preferences from the Display menu, then use the dialog box to set length measurements to cm.

I8. How long does Raul's ride have to be for him to travel through an angle of 2 radians? Use the sketch to show what that angle would look like. Why does this answer make sense?

I9. How long does Raul's ride have to be for him to travel through an angle of 10 radians. Make a prediction, then use the sketch to show what that ride would look like.

Conjecture

Record any insights you had during your investigation. What things that you believe are true in general are based on your specific observations? What patterns did you notice that you cannot explain yet? What questions would you like to explore further?

Follow-up—Exploration 2.4: Raul's Ride

Focus

F1. How many degrees are there in a complete circle? How many radians are there in a complete circle? How many radians are there in one degree? How many degrees are there in one radian?

F2. Write a formula to convert between radian measure and degree measure.

Look Back

L1. In the Focus section of this exploration, you determined the number of radians in a complete circle. *Prove* that your answer is correct.

L2. Prepare a brief presentation for the class, demonstrating and explaining what it means to have an angle whose measure is greater than 360°.

L3. For the carousel problem, which kind of angle measure is more useful—degrees or radians? Explain your choice.

Explore More

E1. When mathematicians, scientists, and engineers graph the sine function, they usually measure angles in radians. Why do you think they do this? After finishing this exploration, look at the sketch Sine Tracer (*Mac*) or Tracer (*Windows*) and compare it to Raul's Ride (*Mac*) or RaulRide (*Windows*).

Teacher's Notes—Exploration 2.4: Raul's Ride

Construct

C1. The speed with which this sketch moves depends upon the speed of your computer's processor.

C2. If you want to stop the ride short of the full distance, click within the sketch to stop the animation.

Investigate

This sketch works by moving point *Raul* along the line at the same speed as point *Raul's Horse* moves around the circle. Because of the way Sketchpad currently computes speed and distance, the points do not move at exactly the same speed. As a result, the arc length traveled by point *Raul's Horse* will not be exactly the same length as the length of the segment along which point *Raul* moves. The difference is extremely small but is noticeable if you use very precise measurements and large distances. You may want to explain this minor discrepancy to students before they begin working on the exploration. You might also simply wait until an anomalous measurement occurs and deal with the problem then.

I1. The lengths may not be exactly the same due to a small difference in animation speed on circles and segments. You may need to explain this to students, or have them skip this step.

I3. Some students may get confused because the answer (90°) seems too obvious.

I6. If students cannot see how to proceed, you may want to encourage them to read the question carefully and set up their sketch so that Raul rides around the circle for the length in question. Students may need to adjust the Preferences settings more than once.

I9. Not all computer screens will accommodate the entire sketch. Students can either adjust the radius or use the scroll bar at the bottom of the window to move among different parts of the sketch.

Conjecture

Students may notice that one radian is always about 57°. They may also comment that a circle of any size always has 2 radians.

Focus

F2. If students refer to a reference text to help them come up with a formula, encourage them to explain *why* their formula works. You may also want to suggest that students consider what a formula for converting between inches and centimeters would look like.

Look Back

L1. You may suggest that students consider the formula for the circumference of a circle.

Explore More

E1. The sketch Sine Tracer (*Mac*) or Tracer (*Windows*) also works well as a class demonstration if you have access to overhead projection equipment.

Section 3: Fundamental Functions

This section has four explorations. Together they introduce students to the six basic functions of trigonometry—sine, cosine, tangent, cotangent, secant, and cosecant. They also help students explore the basic identities that relate these functions to each other.

Exploration 3.1: Area Revisited develops the definition of the cosine function from a unit circle. This exploration expands on the themes and the sketch used in Explorations 1.2 and 1.3. If possible, Exploration 3.1 should be done after Exploration 2.2. In any case, students will need to understand the definition of the sine function to do Exploration 3.1.

Exploration 3.2: How High Is It? develops one physical interpretation of both the tangent function and the relationship between the tangent function and the sine and cosine functions. Students need to have a firm understanding of the unit circle definitions of the sine and cosine functions to do Exploration 3.2. Exploration 3.2 can be done as an independent exploration but if possible should follow Exploration 2.2 and Exploration 3.1.

Exploration 3.3: Reading a Graph uses the graph of the cosecant function to introduce cosecant and secant as the reciprocal functions of sine and cosine. Students need to have a firm understanding of the sine function and its graph to do Exploration 3.3. Exploration 3.3 can be done independently by students who are proficient in Sketchpad, or as a follow-up to Exploration 2.2.

Exploration 3.4: Air on the P String develops physical interpretations on a unit circle for the six basic trigonometric functions. Exploration 3.4 can be done as an independent exploration. It also works well as a demonstration if you have access to overhead projection equipment.

Try the sketch Complete Tracer (*Mac*) or CTracer (*Windows*)—it's one of the many fun and useful supplementary sketches on the accompanying disk.

Exploration 3.1: Area Revisited

One way to start thinking about the sine function is to look at the area of a triangle inscribed in a circle like the one at the right. If you did Explorations 1.2 and 1.3, you saw that if the circle in this diagram has a radius of 1, then the height of the triangle is the sine of the central angle.

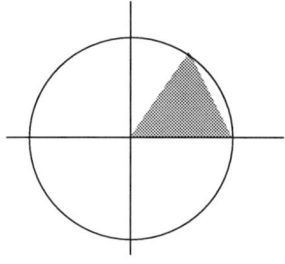

In this exploration, you will investigate the area of a different triangle inscribed in a circle and discover another fundamental function of trigonometry.

Construct

For this exploration, you will need to construct a unit circle centered at the origin of an *x*- and a *y*-axis. In the circle you will need to make a right triangle like the one below.

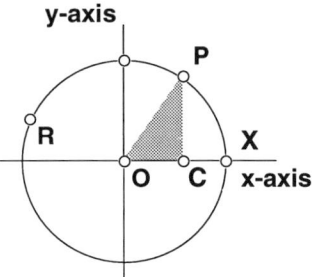

To hide an object, select the object and choose the Hide function from the Display menu.

C1. Draw horizontal line *x-axis* and then line *y-axis* perpendicular to line *x-axis*. Hide their control points.

C2. Construct a medium-sized circle centered at the intersection of the two lines. Label the center point O and the circle *Unit Circle*.

C3. Construct point X at the right-side intersection of circle *Unit Circle* and line *x-axis*.

C4. Construct point P on circle *Unit Circle*. (Don't use the point you used to construct the circle.)

C5. Select point P and line *x-axis*, and construct a perpendicular line.

C6. Construct point C at the intersection of the perpendicular line and line *x-axis*. After you have constructed point C, hide the perpendicular line.

C7. Select points O, P, and C, and construct the polygon interior.

32 • Section 3   ©1995 by Key Curriculum Press   Fundamental Functions

Investigate

I1. Use the Selection Arrow tool to move point *P* around the circle. Describe in qualitative terms how the area of the triangle changes as point *P* moves around the circle.

I2. Measure the central angle of the circle. Make a prediction: At what angle(s) is the area of the triangle the greatest? At what angle(s) is it the least?

I3. Measure the area of the triangle and test your prediction.

I4. Imagine that Sketchpad cannot calculate areas directly. Use Sketchpad to calculate the area of the triangle using its base and height.

If your circle is a unit circle, then the distance from point C to point P is the sine of the central angle.

I5. If the circle is a unit circle, then the distance from point *O* to point *C* is called the **cosine** of the central angle. Write an equation for the area of the triangle using sine and cosine.

I6. What happens to the distance from point *C* to point *P* when the radius of the circle doubles? What happens to the distance from point *O* to point *C* when the radius of a circle doubles? What happens to the area of the triangle when the radius of the circle doubles? Do you need to change your equation to account for this?

 Conjecture

Record any insights you had during your investigation. What things that you believe are true in general are based on your specific observations? What patterns did you notice that you cannot explain yet? What questions would you like to explore further?

Exploring Trigonometry   ©1995 by Key Curriculum Press   The Geometer's Sketchpad • 33

Follow-up—Exploration 3.1: Area Revisited

 Focus

F1. Sketchpad can calculate the sine and cosine of angles for you. To do this, select the *measurement* of the angle you want to use, then choose Calculate on the Measure menu. You calculate the cosine of an angle by choosing Cos[from the pull-down list. Then choose the angle measure from the pull-down menu, and click on). Use this method to have Sketchpad calculate the sine and cosine of the central angle.

F2. Use the values of sine and cosine calculated by Sketchpad to verify the equation for the area of the triangle you wrote in steps I5 and I6.

 Look Back

L1. Based on your knowledge of the sine function, write a precise definition for the cosine of an angle in terms of a unit circle. Make sure your definition works for all central angles of a circle no matter what the radius of the circle is.

L2. You may already know from previous study that the sine function can be defined as the y-coordinate of a point on the unit circle. This means that for some angles, sine is *negative*. Similarly, the cosine function can be defined as the x-coordinate of a point on a unit circle. This means that for some angles cosine is *negative*. For which angles are sine and cosine negative? Does this mean you have to change the equation you wrote and revised in the Investigate section? Explain why or why not.

Explore More

E1. Do you think the graphs of the sine and cosine function will look similar or different? If you have not already done so, make a graph of cosine as a function of the central angle and sine as a function of the central angle and compare.

E2. Based on your observations and/or your graphs of the sine and cosine function, what would a graph of the *area* of the triangle as a function of the central angle look like?

Teacher's Notes—Exploration 3.1: Area Revisited

Investigate

I1. Students often have trouble with qualitative descriptions. Encourage them to explain what they see. An example might be: The area of the triangle increases as the angle goes from 0° to 45°, then decreases as the angle goes from 45° to 90°.

I2. Students sometimes measure angle *COP* as the central angle in the sketch. This will give inaccurate measures for angles greater than or equal to 90°. The central angle should be measured as angle *XOP*.

I6. The distance from point *Origin* to point *C* doubles when the radius doubles because the distance is equal to the cosine of the central angle times the radius. The area of the circle varies with the square of the radius.

Conjecture

Notice that the sine and cosine both have a maximum value equal to 1 because they are both defined within the unit circle. Also, as one increases, the other decreases. They are equal at 45°.

Focus

Some students are tempted to use their own hand calculators to do computations. You may want to encourage students to use Sketchpad's calculator because when they move objects in the sketch, Sketchpad recalculates for them. They can thus test a large number of specific cases easily.

Look Back

L1. Students may refer to the triangle definition of the cosine function. You may want to encourage them to write their description without referring to Adjacent and Hypotenuse. Show them how these terms become unclear for angles greater than 90°. A good answer to this question might be: To find the cosine of an angle using a unit circle, first draw the angle as the central angle of a unit circle. Find the point where the terminal ray of the angle intersects the circle. Make a line through that point perpendicular to the *x*-axis (or, make a vertical line through that point). The distance from the origin (or, the center of the circle) to this vertical line is the cosine of the angle.

L2. This question provides a good opportunity to discuss the reason why trigonometric functions have negative values for some angles.

Explore More

These questions provide an excellent introduction to the properties of the graphs of trigonometric functions. See also Explorations 2.2 and 3.3.

Exploration 3.2: How High Is It?

*Angle of elevation can be measured with an instrument called a **clinometer**. Instructions for making a clinometer from simple materials are in the Explore More section of this investigation.*

One of the most important uses of trigonometry is to measure the height and distance of distant objects when making maps. There are a number of ways to measure the height of objects that are too tall or too difficult to reach. Many methods of measuring height use trigonometry; in fact, one method is to walk a known *distance* from the base of a tall object. Then look up at the top of the object and measure the **angle of elevation**—the angle you have to look up from the ground to see the top of the object.

In this exploration, you will use Sketchpad to find the height of a cliff. You know that when you walk 1 km from the base of the cliff, the top of the cliff can be seen at a 37° angle of elevation.

 Construct

For this exploration, you will need to make a scale drawing of the cliff problem using the information above. Be sure to include the cliff top and bottom, the distance from the base of the cliff, and the angle of elevation.

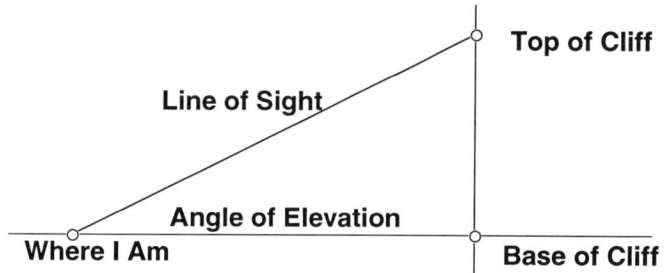

C1. Construct a horizontal line to represent the ground. Hide the control points.

C2. Construct point *Where I Am* and point *Base of Cliff* on the horizontal line.

C3. Select the horizontal line and point *Base of Cliff* and construct a perpendicular line.

C4. Construct point *Top of Cliff* on the perpendicular line.

C5. Construct segment *Line of Sight* between point *Where I Am* and point *Top of Cliff*.

Investigate

I1. This sketch is supposed to be a scale drawing of the situation described at the beginning of this exploration. What is the scale of a drawing and what does that mean about the distances Sketchpad measures?

Sketchpad can measure distances in centimeters. To do this, choose Preferences from the Display menu and set the unit for distance to cm.

I2. Use the Selection Arrow tool to move point *Where I Am* so that it is 10 cm from point *Base of Cliff*. Then construct a circle of radius 10 cm with center at point *Where I Am*. For now let us call this a unit circle. What is our unit?

I3. Measure the angle of elevation, then move point *Top of Cliff* until the angle is 37°.

I4. Use the scale drawing to determine the height of the cliff.

I5. Keep the circle fixed and change the angle of elevation. Describe how the height of the cliff changes as a function of the central angle. For what angle is the height the greatest? For what angle is the height the least?

I6. Construct the sine and cosine of the angle of elevation in your unit circle. Are there similar triangles in your diagram?

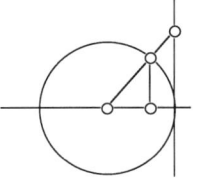

 Conjecture

Record any insights you had during your investigation. What things that you believe are true in general are based on your specific observations? What patterns did you notice that you cannot explain yet? What questions would you like to explore further?

Follow-up—Exploration 3.2: How High Is It?

 Focus

F1. When the circle in the diagram you have drawn is a unit circle, then the height of the cliff is equal to another trigonometric function called the **tangent** of the central angle. Based on your knowledge of the sine and cosine functions, what do you think happens to the relationship between the height of the cliff and the tangent of the central angle when the radius of the circle changes?

F2. Use Sketchpad to calculate the tangent of the angle of elevation in your sketch to confirm your hypothesis about the relationships among the height of the cliff, the radius of the circle, and the tangent of the central angle.

 Look Back

L1. Write an equation that expresses the relationships among the height of the cliff, the distance between the observer and the base of the cliff, and the tangent of the angle of elevation. Be prepared to explain clearly what your equation means in simple English.

L2. At the end of the Investigate section, you constructed some similar triangles. Use these to write an equation for the tangent function in terms of the sine and cosine functions.

 Explore More

Why can you ignore your own height when measuring a very tall object?

E1. Use a protractor to make a simple clinometer like the one at the right using a straw, a piece of string, a small weight, and a few pieces of tape. Find a tall building or other object near your school and use your clinometer to measure its height. Unless the object is *very* tall, you will have to add the height of your eyes to the value you

Exploring Trigonometry ©1995 by Key Curriculum Press The Geometer's Sketchpad • 39

calculate. The angle of elevation you measure with the clinometer is from the level of your eyes—not the level of your feet.

Teacher's Notes—Exploration 3.2: How High Is It?

Construct

Students sometimes prefer to label sketches with whimsical names, or to avoid labels altogether. You may want to encourage them to label sketches such as this one carefully. Clear labeling can make the interpretation of models such as this easier.

Investigate

I1. Students often have a difficult time expressing the idea of *scale* clearly. You may need to offer examples, such as the scale on a map.

I2. The sketch will be more accurate if students construct the circle by selecting point *Where I Am* and point *Base of Cliff* and using Circle By Center+Point from the Construct menu.

I3. This sketch may not fit completely on all computer screens. You may want to suggest that students either change their scale (use 2 cm or 5 cm instead of 10 cm) or use the scroll bars on the right and bottom of the sketch to adjust the sketch.

I5. This can be accomplished by moving point *Top of Cliff*.

Conjecture

Notice that you can adjust for the scale of the drawing by dividing all measurements by the radius of the circle. This may remind students of the definition of the sine and cosine functions.

Focus

F1. This question provides a good opportunity to discuss the similarities between *scale* in a drawing and *proportion* in formulas.

Look Back

These questions ask students to produce functional definitions of the tangent functions. Students may need help formulating equations. Encourage them to experiment, looking for a simple way to express tangent in terms of other things they know.

Explore More

This question makes an excellent project for students. The note in the margin provides a good opportunity to discuss the concept of *significance* in quantities of differing magnitudes. A simple diagram of the situation often helps students see why the height of the eye is important in measuring small heights.

Exploration 3.3: Reading a Graph

As you have probably seen already, trigonometric relationships between angles and lines in triangles and circles can be complicated. They are difficult to describe and difficult to visualize. One of the tools that mathematicians use to understand complex relationships is a **graph**. One of the most important skills a mathematician or a student of mathematics can have is the ability to create, read, and interpret graphs.

In this exploration, you will use graphs to investigate trigonometric functions, and use a graph to discover a new function.

Construct

 Open the sketch Mystery Graph (Mac) or Mystery (Windows).

It may make calculations easier if you make segment Unit Length an even multiple of inches or centimeters.

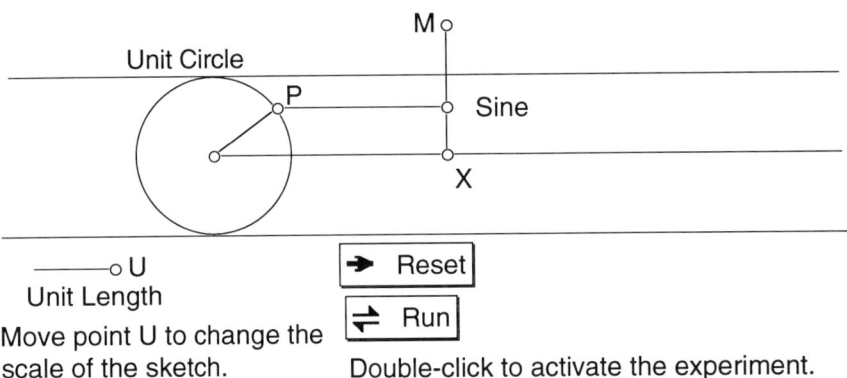

C1. Measure the length of segment *Unit length*.

C2. Use the Selection Arrow tool to move point *U* until the sketch is a convenient size.

Investigate

I1. Double-click on the Reset button. How does this affect point *P* and point *X*?

I2. Double-click on the Run button. Describe what happens in the sketch.

I3. Trace the locuses of point *Sine* and point *M*. To do this, select point *Sine* and point *M* and choose Trace Locus from the Display menu. If point *M* is not in your picture, move point *P* until you can see point *M*.

You can stop the animation anytime by clicking within the sketch. Use this technique to stop the animation and measure distances.

Sketchpad can do the division in I6 for you. Select both measurements, then choose Calculate from the Measure menu.

I4. Reset and run the sketch again. What curve does point *Sine* trace?

I5. Measure the central angle of the circle made by point *P*.

I6. Measure the distance from point *X* to point *Sine*. Divide this value by the length of your unit. Repeat the experiment. Do the values you calculate confirm your observation? Check your hypothesis by comparing the value of the sine to the value you calculate for several angles.

I7. To make your sketch easier to understand, choose the Text tool and double-click on the value you calculated. Label this measure *Scaled Sine*.

I8. Point *M* seems to trace a very different curve. Is this mystery curve a function? Do you think the points on the mystery curve are related to points on the curve made by point *Sine*? Explain your thinking.

You can compare the values of measure Scaled Sine *and measure* Scaled M *by moving point* P *around the circle.*

I9. In the same way you calculated the scaled height of point *Sine* (factoring out the unit length), calculate the scaled height of point *M*. Label the calculated value measure *Scaled M*.

I10. Repeat the experiment. What observations can you make about measure *Scaled Sine* and measure *Scaled M*?

 Conjecture

Record any insights you had during your investigation. What things that you believe are true in general are based on your specific observations? What patterns did you notice that you cannot explain yet? What questions would you like to explore further?

Exploring Trigonometry ©1995 by Key Curriculum Press

Follow-up—Exploration 3.3: Reading a Graph

 **Focus**

F1. Use Sketchpad to calculate measure *Scaled Sine* times measure *Scaled M*. Repeat the experiment and record your observations.

F2. Write an equation for the scaled height of point M using the sine function and the measure of the central angle.

 Look Back

> **Definition:** The **secant** (sec) and **cosecant** (csc) functions are defined as inverses of the sine (sin) and cosine (cos) functions.
>
> $$\sec\theta = \frac{1}{\cos\theta} \quad \text{and} \quad \csc\theta = \frac{1}{\sin\theta}$$

L1. Write the equation for the height of point M in terms of the secant or cosecant of the central angle of the circle in the sketch.

L2. What happens to point M when point P is 180° (π radians) around the circle? Why does this happen?

 Explore More

E1. What would the graph of the secant function look like? Use your knowledge of the graph of the cosine function and your observations in this exploration to sketch graphs of the sine, cosine, secant, and cosecant functions.

E2. In the sketch Mystery Graph (*Mac*) or Mystery (*Windows*), Sketchpad uses animation to trace graphs of two trigonometric functions. Can you create a sketch to trace other functions? Could you use a similar technique to make a tangent tracer?

Teacher's Notes—Exploration 3.3: Reading a Graph

Construct

Students can adjust the scale of the whole sketch by moving point U. They can also control the length of segment *x-axis* by dragging its left endpoint, point A.

Investigate

I5. The simplest way to measure the central angle is to use point A or point X. Students can also construct a point at the intersection of the circle and the ray that has the center of the circle as its endpoint, but that point may make it difficult to move point P later on in the investigation.

I6. Students sometimes have difficulty understanding what it means to confirm an observation. Encourage them to move point P and check their results several times.

I8. Students may be tempted to call the mystery curve a series of parabolas. This provides a good opportunity to discuss the precise definition of a parabola and to refine students' understanding of the graph of a parabola.

I9. Students may have difficulty figuring out how to calculate measure *Scaled M* "in the same way" that they calculated measure *Scaled Sine*. You may want to encourage them to read step I6 carefully and think about how the division they perform is a way of factoring out the unit length.

Conjecture

Notice that the scaled measures vary inversely. Additionally, measure *Scaled Sine* has a maximum value of 1, while measure *Scaled M* has a minimum value of 1.

Focus

F2. If students are having trouble here, you may suggest they rearrange the equation they calculated in step F1.

Look Back

L2. This question provides a good opportunity to discuss the values of the trigonometric functions at quadrantal angles. Exploration 3.4 can help students understand why the values of some of the trigonometric functions are undefined at some angles.

Explore More

These questions provide an excellent transition to the properties of the graphs of the trigonometric functions. See also Explorations 1.2 and 2.1.

Exploration 3.4: Air on the P String

Air on the G String is the title of a famous piece of music by J.S. Bach and the basis for the title of this exploration.

There are many parallels between mathematics and music. One important idea that mathematicians and musicians both cherish is the beauty in **variations on a theme**. In music, a variation on a theme is when two or more pieces of music—or sometimes two or more voices in the same piece of music—use the same basic melody with subtle differences. In mathematics, the same mathematical object or relationship often can be viewed in several different ways. In both cases, looking at the same thing from many perspectives shows us aspects we might not have noticed if we'd always looked at it in the same way.

In this exploration, you will investigate how trigonometric functions can be found in the geometry of a unit circle, and you will discover how the tangent function gets its name.

Construct

For this exploration, you will need to construct a unit circle tangent to the hypotenuse of a right triangle, with its center at the vertex of the right angle. In your construction, point *P*—where the circle is tangent to the hypotenuse—should move freely around the circle, with the other points constructed accordingly.

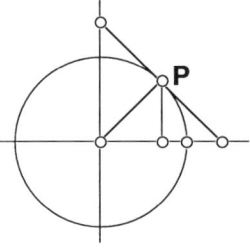

C1. Draw one horizontal line and one vertical line, hide the control points, and construct a point at the intersection of the two lines.

Calculations may be easier if you make the radius of the circle an even multiple of inches or centimeters.

C2. Construct a circle with the intersection point as the center.

C3. Construct points at the intersection of the circle and the lines.

C4. Construct point *P* on the circle (don't use the control point of the circle).

C5. Construct a segment from the center of the circle to point *P*.

C6. Construct a line tangent to the circle at point *P*. To do this, select point *P* and the segment from the center to point *P*, then construct a perpendicular line.

C7. Construct points at the intersection of the tangent line with the horizontal and vertical lines.

C8. Construct a line through point *P* parallel to the vertical line.

C9. Construct a point at the intersection of the parallel line and the horizontal line.

 Investigate

You may have already seen that the six basic functions are usually abbreviated as sin, cos, tan, cot, sec, *and* csc.

Definitions: *There are six basic trigonometric functions:* **sine**, **cosine**, **tangent**, **cotangent**, **secant**, *and* **cosecant**. *The last four functions can all be defined in terms of the first two—that is, in terms of sine and cosine.*

$$\tan\theta = \frac{\sin\theta}{\cos\theta} \qquad \cot\theta = \frac{\cos\theta}{\sin\theta} \qquad \sec\theta = \frac{1}{\cos\theta} \qquad \csc\theta = \frac{1}{\sin\theta}$$

I1. Measure the central angle in your sketch.

If you're having trouble, you might want to think about sine or cosine first. See if you can figure out what the proportion is be-tween the value of the function and the distance in the sketch.

I2. Use Sketchpad to calculate the value of the six basic trigonometric functions for the central angle. To do this, select the measurement of the central angle, then choose Calculate from the Measure menu.

I3. For each of the six basic trigonometric functions, there is a pair of points in the sketch whose distance is proportional to the value of the function. Identify the pair of points for each of the six functions.

 Conjecture

Record any insights you had during your investigation. What things that you believe are true in general are based on your specific observations? What patterns did you notice that you cannot explain yet? What questions would you like to explore further?

Exploring Trigonometry

Follow-up—Exploration 3.4: Air on the P String

 Focus

F1. Based on your observations, where do you think the tangent function gets its name?

F2. Which of the trigonometric functions are *undefined* for some values of the central angle? Use your sketch to explain why this happens.

 Look Back

L1. For each of the six basic functions, use similar triangles to *prove* that the value of the function is proportional to the distance you have identified in your sketch.

L2. Each of the six basic functions are *proportional* to a distance in the sketch. What are the proportions and how could you make the lengths *equal* to the value of the functions, rather than proportional?

 Explore More

E1. Use your sketch to show that $\sec^2 + \csc^2 = (\tan + \cot)^2$. What other relationships among the six basic trigonometric functions can you see in the sketch?

Teacher's Notes—Exploration 3.4: Air on the P String

Construct

It may help to label all of the points in the sketch. It will be easier to refer to the points in the Investigate section.

Investigate

I2. It helps to use the Text tool to rename the measurements that students calculate.

I3. It is useful to discuss, as a class, strategies—including, perhaps, trial and error—for identifying the correct pairs of points.

Conjecture

Notice that the cofunctions (cosine, cotangent, and cosecant) are all mirror images of their respective functions.

Focus

F2. This question provides an excellent opportunity to discuss the meaning of division by zero, infinity, and undefined quantities.

Look Back

L1. It helps to look for a right triangle that contains the radius of the circle, which is 1 in a unit circle, and one of the trigonometric functions in the other. That triangle is usually similar to the right triangle with the radius, sine, and cosine as its sides (look for an angle congruent to the central angle).

Explore More

E1. This question provides an excellent opportunity to discuss the Pythagorean Identity ($\sin^2\theta + \cos^2\theta = 1$) and its many forms.

Section 4: Laws and Their Applications

This section has four explorations. Together they introduce students to the Laws of Sine and Cosine. They also help students use the laws to solve triangles and apply the technique of solving triangles to problems in cartography and navigation.

Exploration 4.1: Navigation introduces the other explorations in this section. Exploration 4.1 uses Sketchpad to solve a complex problem in coastal navigation, but it does not require any formal knowledge of trigonometry. The problem posed in Exploration 4.1 is designed to help students use Sketchpad to model a physical situation. Exploration 4.1 is reviewed in Exploration 4.2—you may want to have students do Exploration 4.1 first.

Exploration 4.2: Invariant Quantities helps students discover the Law of Sines by experiment. In the Focus section they are asked to prove their results geometrically. Students need to have a firm understanding of the sine function and be familiar with Sketchpad to do Exploration 4.2. The Look Back section of Exploration 4.2 reviews the problem from Exploration 4.1. Exploration 4.2 works well as a discovery activity or as a demonstration for the Construct and Investigate sections, followed by individual or group work for the follow-up.

Exploration 4.3: Special and General develops the Law of Cosines and helps students see the Pythagorean Theorem as a special case of the Law of Cosines. Students need to have a firm understanding of the cosine function to do Exploration 4.3. Exploration 4.3 can be done as an independent exploration by students proficient in Sketchpad, or as a follow-up to earlier explorations.

Exploration 4.4: Ambiguity helps students explore the Ambiguous Case of the Law of Sines. Students try to solve a triangle given two sides and an angle not contained between them (Angle-Side-Side). Students work on a problem from navigation and then explore the Ambiguous Case from a theoretical perspective using Sketchpad. Exploration 4.4 can be done independent of other explorations by students proficient in Sketchpad, but it may be easier for students who have already completed Exploration 4.1. Exploration 4.4 is a difficult exploration; except for advanced students, it may work better as a demonstration.

> Try the sketch Law of Cosines (*Mac*) or LawCos (*Windows*)—it's one of the many fun and useful supplementary sketches on the accompanying disk.

Exploration 4.1: Navigation

Trigonometry has many important applications in the real world, but one of the most important is in navigation. Currents, winds, and imperfect instruments make it difficult to travel in precisely straight lines for exact distances. Ships' captains are constantly looking for bearings and elevations to keep from going too far off course—especially near the coast.

In this exploration, you will investigate one way that triangles and trigonometry are used in navigation.

The angle of elevation of an object is the angle you have to look up from the ground (or the surface of the water) to see the top of the object.

> Captain Emily Ahab is sailing from Fisher's Island to Niantic Bay. Her course is tricky because she has to pass through a line of large rock formations known as Bartlett Reef. The fog is thick, and she can just barely see the tops of the rocks above the fog bank.
>
> To pass through the reef, Captain Ahab has to head straight for South Dumpling, the largest rock formation in the whole reef. Then, when she is between 100 and 150 meters from South Dumpling, she has to turn toward Niantic Bay, slipping through the channel between North and South Dumpling with only 50 meters to spare on each side.
>
> Ahab sights the beacon at the top of South Dumpling with a 6° angle of elevation. Six minutes later, she again sights the beacon at the top of South Dumpling, this time with a 15° angle of elevation. At her present speed, Ahab figures she has covered 300 meters in the six minutes between her observations. How much longer should she head for South Dumpling before making her turn?

Construct

For this exploration, you will need to make a scale drawing of the problem using Sketchpad. Start with a line for the surface of the sea, a point for each position of the ship, and a line for the angle of the first sighting of the beacon.

C1. Construct a horizontal line to represent the surface of the ocean. Hide the control points.

C2. Construct point *R* and point *S* on the horizontal line.

C3. Construct a line through point *R* with a 6° angle of elevation. One way to do this is to select point *R* and choose Mark Center from the Transform menu. Then select your horizontal line and choose Rotate from the

Transform Menu. In the dialog box, rotate by a fixed angle of 6°.

 Investigate

I1. Add a line representing the second sighting to your sketch.

I2. Locate the position of the beacon in your sketch, and from that construct point *Rocks* at the base of the rocks below the beacon.

I3. Measure the distance between points R and S.

I4. Measure the distance between points S and *Rocks*.

You may find it convenient to choose a scale that gives you an even factor or multiple of 300 meters.

I5. Choose a scale for your drawing. Adjust the distance between points R and S to fit your scale. What is the distance from the ship to the rocks after the second sighting?

I6. With her present course and speed, how long would it take for Captain Ahab to travel the distance from the second sighting (point S) to the rocks of South Dumpling?

 **Conjecture**

Record any insights you had during your investigation. What things that you believe are true in general are based on your specific observations? What patterns did you notice that you cannot explain yet? What questions would you like to explore further?

Exploring Trigonometry ©1995 by Key Curriculum Press The Geometer's Sketchpad • 53

Follow-up—Exploration 4.1: Navigation

 Focus

F1. How long after the second sighting should Captain Ahab make her turn toward Niantic Bay?

F2. Imagine that a captain of a ship with many years of experience at sea was skeptical about your answer. How would you convince him or her that you had found the correct distance?

 Look Back

L1. In what sense did your solution to the problem depend on your choice of scale?

L2. Can you generalize this technique? Write up an explanation of how you could use this technique to solve other similar problems.

 Explore More

E1. What is Captain Ahab's margin for error? If her sightings are off by as much as 1° in angle of elevation, how far off could she be in estimating the distance to South Dumpling?

E2. Can you think of other examples of real-world problems you could solve by constructing scale models like the one in this exploration? Try to think of an example, and solve the problem using a similar technique.

E3. Can you extend the problem in this investigation to make it more realistic? What other factors might the captain need to consider in a situation such as this one? Try to revise the problem to account for factors such as current, uneven ocean depth, and so on.

You might try to think of another example in navigation. You could also think about problems in architecture, construction, surveying, or cartography.

Teacher's Notes—Exploration 4.1: Navigation

Construct

C3. You may want to encourage students to follow the directions for this step carefully. Rotating the line in the manner described will fix the angle at 6° even when other parts of the sketch are moved in the Investigate section.

Investigate

I1. Students should use the technique from C3.

I5. A good scale might be 100 m = 1 cm.

Conjecture

Notice we are assuming that South Dumpling does not extend into the ocean beyond the beacon at its top. It must therefore be a sheer cliff or another similar formation.

Focus

F1. Students' answers will vary depending on how far from South Dumpling they decide to turn. The range in the problem is 100 to 150 meters. This question provides a good opportunity to discuss the concept of margin of error.

Look Back

L1. Students may regard this as something of a trick question. This question provides an opportunity to discuss the idea that a choice of scale is arbitrary, as long as it is consistent throughout the solution.

Explore More

E1. This question provides a good opportunity to discuss the concept of margin of error.

Exploration 4.2: Invariant Quantities

Invariant quantity is just a fancy, mathematical name for something that is *constant*—that is, it remains the same even while other things change around it. Invariant quantities are important in mathematics because they are useful in solving problems and they often lead us to mathematical rules, which are also known as laws. Things that stay constant are like steady points of light in stormy seas of mathematical uncertainty, leading us to solutions.

In this exploration, you will investigate this simile further, discovering a mathematical law that is often used to solve problems in navigation.

Construct

For this exploration, you will need to construct a triangle from three line segments. Label the vertices of the triangle A, B, and C and the sides opposite the vertices a, b, and c, respectively.

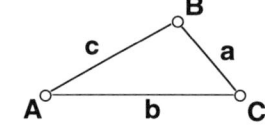

C1. Construct points A, B, and C.

C2. Construct segment a from point B to point C.
Construct segment b from point A to point C.
Construct segment c from point A to point B.

C3. Move point A, point B, and point C so that segment AC is horizontal and point B is above segment AC as shown in the diagram above.

Investigate

I1. Construct the height of the triangle at point B.

I2. Find the sine of the angle at point A using the height you constructed in I1 by measuring line segments and computing the sine from those values. Do not use the sine function on the Sketchpad calculator or another calculator.

I3. Find the sine of angle C *without* making any additional construction lines.

I4. Select the calculated values of sinA and sinC, as well as the length measurements of segment a and segment *c*. Use Sketchpad to calculate the value of the quantity below.

$$\frac{(\text{length of segment } c) \times (\text{sin of angle } A)}{(\text{length of segment } a) \times (\text{sin of angle } C)}$$

15. Use the Selection Arrow tool to change the shape of the triangle. What do you notice?

 Conjecture

Record any insights you had during your investigation. What things that you believe are true in general are based on your specific observations? What patterns did you notice that you cannot explain yet? What questions would you like to explore further?

Follow-up—Exploration 4.2: Invariant Quantities

 Focus

F1. Use your observations to write an equation for $(\sin A/a)$ in terms of c and $\sin C$.

F2. Write up a geometric *proof* that your equation is always true.

F3. Is it also true that $\dfrac{\sin B}{b} = \dfrac{\sin A}{a}$? Explain why or why not.

 Look Back

L1. The equation you wrote in step F1 is more commonly known as the Law of Sines. Is the Law of Sines an example of an invariant quantity? Explain why or why not.

It may be useful to figure out all of the angles in the two triangles first, then use the Law of Sines.

L2. Use the Law of Sines to solve the problem from the last exploration shown below.

> Captain Emily Ahab is sailing from Fisher's Island to Niantic Bay. Her course is tricky because she has to pass through a line of large rock formations known as Bartlett Reef. The fog is thick, and she can just barely see the tops of the rocks above the fog bank.
>
> To pass through the reef, Captain Ahab has to head straight for South Dumpling, the largest rock formation in the whole reef. Then, when she is between 100 and 150 meters from South Dumpling, she has to turn toward Niantic Bay, slipping through the channel between North and South Dumpling with only 50 meters to spare on each side.
>
> Ahab sights the beacon at the top of South Dumpling with a 6° angle of elevation. Six minutes later, she again sights the beacon at the top of South Dumpling, this time with a 15° angle of elevation. At her present speed, Ahab figures she has covered 300 meters in the six minutes between her observations. How much longer should she head for South Dumpling before making her turn?

L3. Is your solution from the last exploration using Sketchpad *more* or *less* convincing than your solution using the Law of Sines? Which method is more practical? Explain the reasons behind your answers.

Explore More

E1. What kinds of triangles can you solve using the Law of Sines? The Law of Sines can help you figure out *all* the sides and angles of a triangle when you only have *some* of them to begin with. What do you need to know about a triangle to use the Law of Sines? See if you can list all possible cases. In what situations does the Law of Sines *not* work?

E2. Think of an example of a real-world problem you could solve using the Law of Sines, then solve it.

Teacher's Notes—Exploration 4-2: Invariant Quantities

Construct

C3. The Investigate section is easier to follow if students construct their triangle as shown in the diagram, with segment AC horizontal and point B above segment AC.

Investigate

I1. Some students may try to construct the height by drawing a segment. This will work until they move one of the points of the triangle. They should construct a line perpendicular to segment AC through point B.

Conjecture

Students may notice that the computed value remains constant while point B is above segment AC. When the height disappears, so does the computed value. A good conjecture might be that the disappearance is a consequence of the way Sketchpad is computing sinA and sinC. If you measure angle A and angle C and use Sketchpad's calculator to compute the sines from angle A and angle C directly, the value should remain constant no matter where point B moves in the sketch.

Focus

F2. The proof follows from the diagram used in the Investigation section. You may need to suggest to students that they rewrite their equation from F1 by substituting opposite/hypotenuse for the sine functions in the equation.

Look Back

L1. Technically speaking, the Law of Sines is *not* an invariant quantity. Both quantities in the equation change as the dimensions of the triangle change. At the same time, the Law of Sines can be expressed as an invariant quantity by rewriting it in the form given in step I4. Interestingly, all mathematical "laws" can be expressed as invariant quantities in the same way. For instance, the Pythagorean Theorem is equivalent to the invariant: $(a^2 + b^2)/c^2 = 1$

L2. This is a difficult problem, mostly because of the amount of information presented. You may want to suggest that students draw a careful diagram, then solve the triangles they draw using the Law of Sines.

Explore More

E1. This question presents an excellent opportunity to discuss some of the limitations of the Law of Sines. Students should notice that the Law of Sines requires three pieces of information about a triangle, including one known angle and the length of a side opposite it.

Exploration 4.3: Special and General

New mathematical ideas are often discovered by looking at *specific* theorems or situations and trying to discover a *general* rule that explains them. Specific mathematical examples often come out of questions that people encounter in the real world, or in the methods they use to solve common problems. General theorems are not always as useful as specific solutions, but they are sometimes more *beautiful* because they show how things that appear different are examples of the same basic principle.

In this exploration, you will investigate the Law of Cosines and discover that it is a generalization of one of the most famous theorems in geometry.

Construct

For this exploration, you will need to construct a triangle from three line segments. Label the vertices of the triangle A, B, and C, and the sides opposite the vertices a, b, and c, respectively.

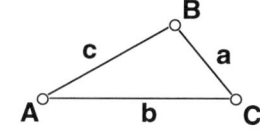

It will make calculations easier if you re-label your measurements. Choose the Text tool and double-click on a measurement. Use the dialog box to shorten the label of the measure.

C1. Construct triangle *ABC*.

C2. Measure the length of all the sides. Measure all the angles. Make sure that Sketchpad reports positive angle measures. If it does not, choose Preferences from the Display menu and set Angle Measure to degrees.

Investigate

I1. In this sketch, you can move all three of the vertices of the triangle. For now, leave point *A* and point *B* where they are, and move point *C* around using the Selection Arrow tool. What measures stay the same? What measures change?

Select the measurements of segment a and segment b, then choose Calculate from the Measure menu.

I2. Use Sketchpad to calculate $a^2 + b^2$. What happens to this quantity as you move point *C*?

I3. Use Sketchpad to calculate cos*C*. What happens to this quantity as you move point *C*?

Use the Text tool to relabel your calculations.

I4. Use Sketchpad to calculate $-2ab\cos C$. What happens to this quantity as you move point *C*?

15. Sketchpad has calculated $a^2 + b^2$ and $-2ab\cos C$ for you. Select these two calculations and use Sketchpad's calculator to add them together. What happens to the quantity $a^2 + b^2 - 2ab\cos C$ as you move point C?

16. Move point C so that the angle of the triangle at point C is $90°$. Record your observations.

17. Move point A. Then move point C to various locations. Record your observations.

18. Use Sketchpad to calculate c^2. Repeat 17. Record your observations.

 Conjecture

Record any insights you had during your investigation. What things that you believe are true in general are based on your specific observations? What patterns did you notice that you cannot explain yet? What questions would you like to explore further?

Follow-up—Exploration 4.3: Special and General

Focus

F1. What quantities are always equal in the sketch? Write an equation to express this.

F2. Explain why the Pythagorean Theorem is a special case of the equation you just wrote.

Look Back

L1. The equation that relates c^2, $a^2 + b^2$, and $-2ab\cos C$ is usually known as the Law of Cosines. Write a similar equation using $\cos A$. Write up a clear explanation of the reasoning behind your answer.

L2. The Law of Cosines is often used to *solve* a triangle—to figure out *all* the sides and angles of a triangle when you only have *some* of them to begin with. What do you need to know about a triangle to use the Law of Cosines to figure out all of the sides and angles of a triangle?

Explore More

E1. Can you make up a real-world problem that you would use the Law of Cosines to solve? What *other* methods could you use to solve the same problem? Which method do you think would be the best method to use?

Teacher's Notes—Exploration 4.3: Special and General

Construct

C2. The screen will quickly become cluttered with measurements in this exploration if students do not relabel their measurements and calculations.

Investigate

In several of the steps of the Investigate section, students are asked to observe what happens to a measured or calculated quantity as they move points in the sketch. In many cases, the quantity changes as the point moves, with no easily discernible pattern. Some students may find this frustrating. This may be a good opportunity to discuss the reason that invariant quantities are so useful, so valued, and so rare, in mathematics.

Conjecture

Many students will notice that $a^2 + b^2$ and $-2ab\cos C$ is always the same as c^2.

Focus

F2. If students are having trouble with this question, you might suggest that they repeat step I6 and consider what happens to their equation from F1 when angle C is 90°.

Look Back

L2. This question provides a good opportunity to compare and contrast the Law of Sines and Law of Cosines. See also Exploration 4.2, question E1.

Explore More

E1. Many textbooks provide examples of such problems.

Exploration 4.4: Ambiguity

As you may have already seen, the Law of Sines is a powerful tool for solving practical problems involving angles and distances. It is easy to remember, easy to use, and it gives us a great deal of information about a triangle. But there are times when the Law of Sines is deceptively simple.

In this exploration you will investigate cases where the Law of Sines gives ambiguous (uncertain) results, and you will discover guidelines for using this powerful tool with certainty.

> The ship *Captain Bligh* is on a research trip with the submarine *Intrepid*. Both craft start at the same location off the coast of Virginia. The *Intrepid* submerges and heads due east. The *Captain Bligh* sets out on a course 30° north of the *Intrepid*.
>
> When the *Captain Bligh* has traveled 5 km, the radio operator receives a distress call from the *Intrepid*. The submarine's engines have stopped, and it is trapped below the surface. Based on the distress signal, the radio operator estimates that the distance to the submarine is 4 km, but there is too much interference for her to determine the exact direction to the *Intrepid*.
>
> The captain of the ship needs to know how to get to the submarine. Use the Law of Sines to determine the direction from the *Captain Bligh* to the *Intrepid*.

After you have tried solving the problem using the Law of Sines, start the exploration below to help you understand the problem better.

Construct

For this exploration, you will need to construct a scale drawing of the problem.

C1. Construct point *Old Position*.

C2. Construct horizontal ray *Sub's Route* starting at point *Old Position*.

If 5 cm represents 5 km, what is the scale of the drawing?

C3. Construct point *New Position* by selecting point *Old Position* and translating it by Fixed Polar 5 cm at 30°. To translate the point, choose Translate from the Transform menu.

C4. Construct segment *5 km @ 30 deg* between point *Old Position* and point *New Position*.

C5. Construct segment *Dist. to Sub*. Measure the segment and move its endpoints until it is 4 cm long.

Investigate

I1. Construct a circle with center at point *New Position* and radius equal to segment *Dist. to Sub* by selecting the point and the segment and choosing Circle by Center + Radius from the Construct menu. What does this circle represent in terms of the original problem?

I2. Select the circle and ray *Sub's Route*. Choose Point at Intersection from the Construct menu. This will create two points. What do these points represent in terms of the original problem?

I3. If you have not already done so, use the Law of Sines to verify both possible solutions to the Ship/Submarine problem. What important number do the two solutions have in common?

 Open the sketch AngleSideSide *(Mac) or* AngSS *(Windows).*

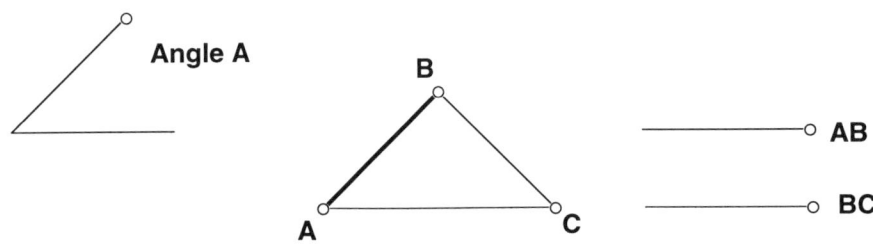

In the sketch AngleSideSide *(Mac) or* AngSS *(Windows), point* AB *controls the length of segment* AB. *point* BC *controls the length of segment* BC, *and point Angle A controls the size of angle with vertex point* A.

I4. Move point *AB*. Do an angle and two sides always define a triangle?

I5. Try moving point *AB*, point *BC*, and point *Angle A*. Record your observations.

I6. Construct the height of the triangle at point *B* by selecting the horizontal ray and point *B*, then choose Perpendicular Line from the Construct menu. How could you calculate the length of this segment in terms of angle *A* and segment *AB*?

I7. Does this height give you any insights into the conditions that let an angle and two sides like those shown in sketch AngleSideSide (*Mac*) or AngSS (*Windows*) define exactly one triangle?

 Conjecture

Record any insights you had during your investigation. What things that you believe are true in general are based on your specific observations? What patterns did you notice that you cannot explain yet? What questions would you like to explore further?

Follow-up—Exploration 4.4: Ambiguity

 Focus

F1. What must be true of segment *AB* and segment *BC* if an angle and two sides like those shown in the sketch define exactly one triangle?

F2. What must be true of segment *AB*, sin*A*, and segment *BC* if an angle and two sides like those shown in the sketch define exactly two triangles?

F3. What must be true of segment *AB*, sin*A*, and segment *BC* if an angle and two sides like those shown in the sketch define exactly zero triangles?

 Look Back

L1. What important geometric theorem (or *nontheorem*) could you use to explain your observations in this exploration?

L2. Prepare a brief presentation for the class explaining what problems can arise when using the Law of Sines to determine information about certain kinds of triangles. Include in your presentation a method or methods for avoiding these problems *without* Sketchpad.

 Explore More

E1. Can you make up problems like the Ship/Sub problem that represent other ways the Law of Sines can malfunction when you are trying to determine information about a triangle?

Teacher's Notes—Exploration 4.4: Ambiguity

Construct

You may want to emphasize to students the importance of clear labeling in complex problem situations such as the problem at the beginning of this exploration.

Investigate

Some students may find this exploration long and difficult. One way to simplify the exploration is to skip the Construct section and the first three steps of the Investigate section. The exploration is designed to function if students start with I4.

I3. This step may require considerable work by students.

I4. This question provides a good opportunity to discuss the idea of systematic investigation. Some students may be tempted to move the controlling points in a haphazard manner. You may want to suggest that they move them one at a time and try to find patterns in their observations.

Conjecture

Notice that when segment *BC* is shorter than the height constructed in I6, the triangles disappear. Similarly, when segment *BC* is longer than segment *AB*, there is only one triangle.

Focus

Some students may report answers in terms of the height they construct at point *B*. You may want to insist that they report this height in terms of segment *AB* and sin*A*. It is equal to (*AB*)sin*A*.

Look Back

L1. The title of the sketch should be a good hint for students.

Explore More

E1. Textbooks may provide additional sample problems for students.

Section 5: Investigation—Motion of a Pendulum

This section has four explorations. Together they make up a single extended investigation into the physics and mathematics of the motion of a pendulum. The extended investigation is designed to help students apply their knowledge of trigonometry in a meaningful way so as to understand the world around them. Because these explorations form a single extended investigation, it would be difficult to do them individually or in a different sequence than they are presented here.

Exploration 5.1: Building a Pendulum asks students to construct two pendulums and explore their periods in various ways. This exploration does not use Sketchpad or require any prior knowledge of trigonometry. You will need to provide string and some small weights (washers from a hardware store work nicely).

Exploration 5.2: Modeling a Pendulum helps students create a model of the motion of a pendulum using Sketchpad. Students need to be familiar with the sine function to complete this exploration. If you are pressed for time and have access to overhead projection equipment, you might consider doing all or part of Exploration 5.2 as a demonstration.

Exploration 5.3: Gravity is the most difficult exploration of the extended investigation. This exploration helps students construct a basic force diagram for the pendulum. Students then explore the way the force of gravity affects the pendulum at different points along its path of motion. Students need to have a firm understanding of the sine function to complete Exploration 5.3. Knowledge of vectors is useful, but students can complete the exploration without it.

Exploration 5.4: Simple Harmonic Motion completes the extended investigation, asking students to use their observations to construct an equation for the period of a pendulum and to compare the equation to their experimental results from Exploration 5.1. Students do not need to have any additional knowledge of physics or trigonometry to complete this exploration, although they do need to have finished Explorations 5.1 through 5.3.

Try the sketch Pendulum (*Mac*) or Pndulum (*Windows*)—it's one of the many fun and useful supplementary sketches on the accompanying disk.

Exploration 5.1: Building a Pendulum

Pendulums have been used by mathematicians, physicists, navigators, astronomers, and clock makers for hundreds of years because of their consistent and predictable motion. Over 100 years ago, a giant pendulum was used to measure the motion of the earth. Pendulums are still used today to keep time in some clocks.

In this exploration you will investigate the factors that affect the motion of a pendulum.

Construct

You will not be using Sketchpad for this exploration.

C1. Build a pendulum using a string 2 to 3 feet long and one weight. Both of these items will be provided by your teacher. To make a simple pendulum, tie the weight to one end of the string. Hold the other end of the string and allow the weight to swing freely. You may want to tie the free end of the string to a pencil or stick to make it easier to hold.

C2. Build a second pendulum, the same length as the first pendulum, but with *two* weights on the end. If you decide to tie the second pendulum to the same pencil or stick you used for the first pendulum, make sure they can both swing freely.

Investigate

> **Definition**: The **period** of a pendulum is the time it takes to travel from the highest point of its swing on one side to the highest point on the other side and then all the way back again. That is, the period is the time it takes to make one complete swing.

I1. Does the number of weights on the end of the pendulum affect its period? Find out using your pendulums and record your results.

You may want to save your pendulums to use again if you are going to do more explorations in this section.

I2. Design and conduct an experiment to discover if the length of a pendulum affects its period.

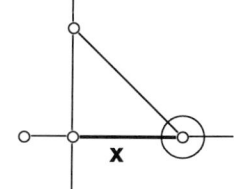

Definition: The **x-displacement** of a pendulum is the horizontal distance of the weight from a vertical line through the vertex (the knot at the top).

I3. How does the *x*-displacement of a pendulum *when you first let go of the string* affect the period of a pendulum? Find out using your pendulums and record your results.

 Conjecture

Record any insights you had during your investigation. What things that you believe are true in general are based on your specific observations? What patterns did you notice that you cannot explain yet? What questions would you like to explore further?

Exploring Trigonometry

Follow-up—Exploration 5.1: Building a Pendulum

Focus

F1. What factors appear to influence the period of a pendulum? Justify your statement.

F2. What factors do not appear to influence its period? Justify your statement.

Look Back

L1. The effect of which factors are the most difficult to observe? The easiest? Try to explain why some things were easier to measure and evaluate than others.

L2. How could you redesign your experiments so that they would be easier to conduct and be more convincing to a skeptical observer?

Explore More

E1. Are there other factors about the construction of the pendulum that might affect its period? Does the thickness of the string or the shape of the weight change the period of a pendulum? Design a series of experiments to see how these or other factors affect the period of the pendulum.

E2. How do you think the period of a pendulum on the moon would compare to the period of that same pendulum on Earth? Explain why you think the way you do.

Teacher's Notes—Exploration 5.1: Building a Pendulum

Construct

Most students have little trouble constructing the required pendulums. Occasionally they swing the pendulum by holding it at the top and moving their hand. The equation for motion of a pendulum assumes that the vertex (knot at the top) is stationary.

Investigate

The most common difficulty students experience in this section is deciding whether the effects of weight or initial x-displacement are significant. This provides a good opportunity to discuss margin of error. It is also an excellent opportunity to discuss the need for interpretation in the conduct of scientific experiment. Most students realize by the end of the investigation that weight and initial x-displacement are not significant *in comparison to* the length of the string. If students have trouble seeing this, you may want to help them; in any case their misconceptions will be cleared up by the end of Exploration 5.4.

I2. Some students need help understanding the definition of the period. You may want to demonstrate by holding a pendulum and letting it complete exactly one swing.

I3. Some students have difficulty understanding the concept of x-displacement. You may need to draw a diagram for them or demonstrate by holding a pendulum on a blackboard and drawing in the x-displacement. Once they understand x-displacement, some students still have difficulty with the concept of *initial* x-displacement. You may need to explain that this refers to the x-displacement of the pendulum before they release it. You may rephrase the question in this way: Does it change the period of the pendulum if I let it go from a different height?

Conjecture

Students will probably summarize the results of their experiments. They may ask why some of these factors affect the pendulum while others do not. They may also express surprise that weight does not seem to change the period.

Focus

Students often do not answer these questions in the way you expect for reasons discussed above (see Investigate). Students may ask you to check their work. You may want to ask them to decide for themselves and proceed with the investigation.

Look Back

These questions provide a good opportunity to discuss the process of systematic investigation.

Explore More

This question makes a good extended project for interested students.

Exploration 5.2: Modeling a Pendulum

Architects use models of buildings to understand their designs before construction starts. Museums use models to show what something looks like when the real object is not available or is difficult to see. Can you think of other ways people use models?

What is a *mathematical* model? A **model** is something that looks or acts in some important way like an object or process in the real world. We experiment with a model because it is easier (and sometimes safer) than using the actual object we want to study; we then use our understanding of the model to explain what is happening in the real world. For example, engineers test aircraft wings by putting models of new designs into a wind tunnel to see how well they perform. This is much less expensive and far less dangerous than initially testing new wings on a plane 10,000 feet in the air!

A **mathematical model** is simply a model that uses the tools of mathematics. A mathematical model can be a set of equations, or a computer program, or a Sketchpad sketch. The important thing is that the mathematical objects in the model behave in some important way like the objects they are modeling.

In this exploration, you will use Sketchpad to make a mathematical model of your pendulum. Later, you will use that model to help you understand the motion of pendulums.

Construct

Don't try to animate your model of the pendulum yet. It will be easier to do after you have investigated the motion of the pendulum further.

C1. On a piece of paper, draw the path of the weight of a pendulum as it swings.

C2. Use Sketchpad to make a simple sketch of the path of the weight of a pendulum. Include in your sketch the path, the weight, the string, and the vertex (the knot at the top) of the pendulum.

Investigate

When you finish constructing and checking x, you may want to save your sketch to use in the next exploration.

I1. Construct a line segment in your sketch that shows the x-displacement of the pendulum. Label this segment x. Make sure that segment x shows the x-displacement correctly, no matter where the weight of the pendulum is on its path.

I2. Choose an angle A in your sketch so that x is proportional to $\sin A$. Two things are **proportional** if they change at the same rate—when one doubles, the other doubles; when one becomes one-third as big, the other becomes one-third as big. One way to make lines proportional is to use similar triangles.

76 • Section 5 ©1995 by Key Curriculum Press Motion of a Pendulum

I3. Measure angle A and the length of segment x. Use these values to calculate $(x/\sin A)$, which is the ratio of x to $\sin A$.

I4. Verify that $\sin A$ is proportional to x by moving the weight of the pendulum to several positions and checking that the ratio you calculated is constant.

 Conjecture

Record any insights you had during your investigation. What things that you believe are true in general are based on your specific observations? What patterns did you notice that you cannot explain yet? What questions would you like to explore further?

Follow-up—Exploration 5.2: Modeling a Pendulum

Focus

F1. Write a true equation that uses x, $\sin A$, and the length of the pendulum.

F2. Prove your equation is true.

Look Back

L1. Explain how your sketch and equation model the motion of the pendulum.

Explore More

E1. Can you construct a sketch that uses animation to model the actual motion of a pendulum? You can get the pendulum to slow down as it reaches the ends of its swing by constructing a sketch like the one shown at the right and animating point D on the circle.

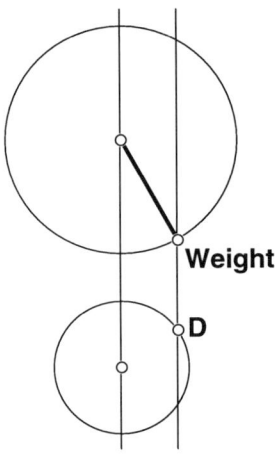

Teacher's Notes—Exploration 5.2: Modeling a Pendulum

Construct

C2. The simplest and most effective model is a circle. The center of the circle is the vertex (knot at the top) of the pendulum. Put a point on the circle to represent the weight, and a segment to represent the string.

Investigate

Students are reminded in the margin to save clean copies of their model to use in the next exploration. You may want to remind them again. Some students have difficulty understanding the concept of *x*-displacement. You may need to draw a diagram for them or demonstrate by holding a pendulum on a blackboard and drawing in the *x*-displacement. Once they understand *x*-displacement, some students still have difficulty with the concept of *initial x*-displacement. You may need to explain that this refers to the *x*-displacement of the pendulum before they release it. You may rephrase the question in this way: Does it change the period of the pendulum if I let it go from a different height?

I1. Students may try to draw segment *x* by hand. This will not show the *x*-displacement correctly when the weight of the pendulum moves. They should construct a vertical line through the center of the circle, then construct a line perpendicular to the vertical through the point that represents the weight of the pendulum.

I2. The easiest angle to choose is the angle at the center of the circle formed by the string of the pendulum and a vertical line. The sine of this angle is proportional to segment *x* because segment *x* is the opposite side of a right triangle.

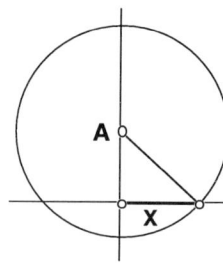

I3 When students measure the central angle of the circle, they should measure using the "weight" of the pendulum, the center of the circle, and the point at the bottom of the circle where the vertical line intersects the circle. If they measure using the point on the vertical line that defines the *x*-displacement, the measure will disappear at 90 degrees.

Conjecture

Notice that this model works only when the weight of the pendulum is below the center of the circle. Also, students may observe that (sin$A$ = x/radius).

Focus

F2. Students may verify by example using Sketchpad. A more formal proof comes from the diagram that shows a right triangle with segment *x* opposite angle *A* and the length of the pendulum as the hypotenuse.

Look Back

L1. This question provides a good opportunity to discuss the concept of a mathematical model. Some students may skip the introduction to the exploration and start with the Construct section. You may need to suggest that they reread the introduction carefully before answering this question.

Explore More

E1. This question provides a good introduction to using the animate feature of Sketchpad. To animate the point, select the point and the circle. Then choose Animate from the Action Button submenu in the Edit menu.

Exploration 5.3: Gravity

You may have noticed in your earlier explorations that when you lift up the weight of the pendulum and let go, something pulls the pendulum down. The "something" is gravity. Gravity is a force of attraction between all objects in the universe. Gravity pulls us toward the earth; it is what keeps us from flying off into space. Gravity keeps the moon in orbit around the earth, and the earth and other planets in orbit around the sun. But if gravity is pulling *down*, why does the pendulum move in a *circle*?

In this exploration you will use your model of a pendulum to help you understand how the pull of gravity affects the pendulum.

 ### Construct

 Open the sketch you saved from the last exploration. If you did not save a sketch, open the sketch Pendulum Model (Mac) or Pendulum (Windows).

 ### Investigate

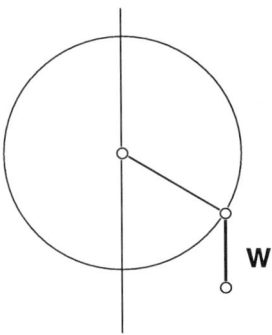

In most pendulums, the weight is much heavier than the string, so to make our model simpler, we assume that gravity is pulling only on the weight.

I1. Construct a segment representing the force (the pull) of gravity. The line should be straight down from the weight of the pendulum, no matter where the weight moves on the path of the pendulum.

I2. Label the segment W. The length of this segment will represent how strong the force of gravity is.

When the force of gravity is perfectly tangent to the circle, all of the force is working to pull the weight around the circle

I3. Move the weight of the pendulum so that angle *A* is equal to 90°. How much of the force of gravity (segment W) is pulling *tangent* to the circle?

I4. Make angle *A* equal to 0°. Now how much of the force of gravity is pulling *tangent* to the circle?

I5. Hold your pendulum (a real one, not the Sketchpad model!) so that angle *A* is 0°. What happens when you let go? How does this relate to your Sketchpad model?

You will probably want to try a few very small angles, a few angles near 45°, and a few angles near 90°.

I6. Repeat the last two steps for different measures of angle *A*. Then describe how the force of gravity pulling *tangent* to the circle changes as angle *A* changes.

I7. Construct a line tangent to the circle at the point that represents the weight of the pendulum.

I8. Make a segment that shows *how far* along the tangent line the force of gravity pulls.

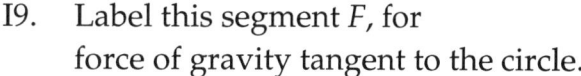

I9. Label this segment *F*, for force of gravity tangent to the circle.

I10. How does the length of segment *F* relate to sin*A*?

Conjecture

Record any insights you had during your investigation. What things that you believe are true in general are based on your specific observations? What patterns did you notice that you cannot explain yet? What questions would you like to explore further?

Follow-up—Exploration 5.3: Gravity

 Focus

> **Definitions**: *In physics, the letter* **m** *usually stands for the* **mass** *of an object, which technically means how hard it is to move. The letter* **g** *usually stands for a constant (unchanging) number that tells us how strong the pull of* **gravity** *is on objects at a particular place. The* force of gravity *on an object—more commonly called an object's* weight—*is equal to* m *times* g, *or* mg.

F1. In your sketch, let us assume that the length of segment W (the length that represents the force of gravity) is *mg*. Based on this assumption, write a true equation involving sinA, *mg*, and F.

F2. Prove that your equation is correct.

 Look Back

You might want to think about what W is doing when angle A is 0°.

L1. If segment F in your model represents the part of W (the force of gravity) that pulls the weight around the circle, where is the *rest* of W, and what is it doing?

L2. Why is it that in the Focus section we *assume* that the length of segment W is equal to *mg*? Do we *know* that segment W is the same as *mg*?

Explore More

E1. Based on your observations, what would a graph of angle A versus segment F look like? What would a graph of sinA versus segment F look like? What would a graph of segment *x* versus segment F look like? Check your answers by creating the graphs.

Teacher's Notes—Exploration 5.3: Gravity

Construct

There is no Construction section for this sketch. If students are not proficient with Sketchpad, you may want to do I1 through I8 as a demonstration.

Investigate

I1. Segment W needs to be constructed by first making a line through the weight of the pendulum parallel to a vertical line through the center of the circle. Students can then construct a point below the weight, hide the parallel line, and construct segment W.

I3. Students may have trouble understanding what it means for part of a force to be pulling on an object. You might ask them to think about a railroad car on a track. If you pull the car *along* the track, all of your pull helps the car move forward. If you pull from the side, none of your pull helps the car move forward.

I6. Students often have trouble providing a general description of how quantities change. This may be a good answer: "The bigger angle A is, the more gravity pulls tangent to the circle."

I7. To make a tangent line, construct a line perpendicular to the segment that represents the string through the point that represents the weight.

I8. Students sometimes have trouble constructing segment F. The key is to construct a line through the bottom point of segment W perpendicular to the line tangent to the circle.

Conjecture

Notice that the right triangle containing segment F and segment W also contains an angle congruent to angle A (the central angle of the circle). This means that there are two similar right triangles in the sketch.

Focus

F2. Students may verify by example using Sketchpad. A formal proof follows from the fact that segment W is parallel to the vertical line through the center of the circle. This can be used to prove that the two right triangles in the sketch are similar.

Look Back

L1. A demonstration is often effective here. Holding a pendulum at rest and asking why gravity does not move the pendulum often helps students see that the string opposes some of the force of gravity.

L2. This is an interesting philosophical question that provides a good opportunity to discuss the meaning of the constant *g* and the concept of scale in a model.

Explore More

E1. This question provides a good opportunity to discuss the value of graphs for understanding functional relationships.

Exploration 5.4: Simple Harmonic Motion

As you have seen, a pendulum oscillates (moves back and forth) in an even way due to the force of gravity. In the study of physics, this kind of even oscillation is called *simple harmonic motion*. All objects that exhibit this kind of motion can be described with the same set of mathematical equations.

In this exploration you will use the general equations of simple harmonic motion to explore the motion of your pendulum.

Construct

You will not be using Sketchpad for the first part of this exploration. You will need to use Sketchpad in the Look Back and Explore More sections.

You will also need to refer to the Focus section of Exploration 5.1 later in this investigation.

C1. In the Focus section of Exploration 5.2, you wrote an equation using x and sin$A$. Copy that equation onto a sheet of paper.

C2. In the Focus section of Exploration 5.3, you wrote an equation involving sin$A$, mg, and F. Copy that equation onto the same sheet of paper.

Investigate

I1. Sine is a very important function, but it is sometimes difficult to use—more difficult than multiplication, for instance. Based on the two equations you have written down, come up with a *new* equation that uses F, m, g, and x (and maybe some other things too), but *not* sin$A$.

I2. Rewrite your new equation so that it is in the form F = _____.

> **Definition**: A **constant** *is a quantity that does not change during an experiment.*

I3. Your new equation shows that F is equal to x times a constant. Explain what this means and how your equation shows that this is true.

> **Definition**: An object whose motion obeys an equation of the form force = constant × displacement *is said to be in* **simple harmonic motion**. Objects in simple harmonic motion oscillate back and forth in a repeating pattern.

I4. According to the definition above, does a pendulum move in simple harmonic motion? Explain your answer.

I5. The period of an object in simple harmonic motion is always given by the formula: period = $2\pi\sqrt{m/k}$, where m is the mass of the object, and k is the constant from the equation force = constant × displacement. Use this formula to write an equation for the period of a pendulum.

> **Definition**: A **parameter** is a quantity in an equation that does not change during an experiment but which can change from one experiment to the next. For instance, the mass of the pendulum is a parameter because you can experiment with the pendulum, change its mass, and experiment again.

I6. Simplify your equation for the period of a pendulum. Which of the remaining terms are constants? Which of the remaining terms are parameters?

 Conjecture

Record any insights you had during your investigation. What things that you believe are true in general are based on your specific observations? What patterns did you notice that you cannot explain yet? What questions would you like to explore further?

Follow-up—Exploration 5.4: Simple Harmonic Motion

Focus

F1. According to your equation, what factors affect the period of a pendulum?

F2. How does the equation you wrote for the period of a pendulum compare to your conclusions in the Focus section of Exploration 5.1?

Look Back

L1. Prepare a presentation that explains in simple terms what your equation for the period of a pendulum *means*. Imagine that someone in your class got lost in the mathematics of the derivation. How would you explain the equation to him or her?

L2. What would a graph of F versus x look like? Why would this be true?

Explore More

E1. What additional information and equipment would you need to verify your equation for the period of a pendulum experimentally? How could you get the information you are missing? Find the equipment and information you need and design an experiment to verify your equation using data from the pendulums you constructed in Exploration 5.1.

E2. Can you design an experiment to verify that F is equal to x times a constant for a physical pendulum (as opposed to your Sketchpad model)?

Teacher's Notes—Exploration 5.4: Simple Harmonic Motion

Construct

This exploration does not require Sketchpad.

Investigate

This section provides an excellent opportunity to discuss the similarities and differences among the terms constant, parameter, and variable.

Conjecture

Notice that the only parameter in the final equation is the length of the pendulum.

Focus

Students may have difficulty seeing that if their equation is correct, then the mass and initial x-displacement of the pendulum cannot affect its period. You may need to explain that if these factors *did* affect the period, then the equation would give inconsistent (and therefore inaccurate) results.

Look Back

L2. This question provides a good opportunity to discuss why quantities that are directly proportional have graphs that are straight lines.

Explore More

These questions provide a good opportunity for students to explore the connections between mathematics and physics. Some rulers and a spring balance may be all that is needed to verify some important parts of the original model.

Section 6: Introduction to Polar Coordinates

This section has three explorations. Together they introduce students to a system of plotting points in a plane known as polar coordinates. These explorations also help students to see the connections between polar and Cartesian coordinates, as well as to graph simple polar equations.

Exploration 6.1: Cartesian Coordinates reviews the definition of Cartesian coordinates and introduces students to some of the ways Sketchpad can represent a coordinate system in the sketch plane. Exploration 6.1 does not require any knowledge of trigonometry; students familiar with Sketchpad can do this as an independent exploration.

Exploration 6.2: Polar Coordinates asks students the same kinds of questions that appear in Exploration 6.1, this time using polar coordinates. Exploration 6.2 introduces students to the definition of *r* and theta coordinates, as well as graphs in a polar coordinate system. This exploration does not assume any specific knowledge of trigonometry, other than the ability to understand positive and negative angle measure. Students familiar with Sketchpad can do the exploration independently, although they may get more from it if it follows Exploration 5.1.

Exploration 6.3: Trigonometry and Coordinate Systems helps students derive equations to convert between polar and Cartesian coordinates. Exploration 6.3 requires a solid understanding of the sine and cosine functions, as well as familiarity with polar coordinates. Students who understand basic trigonometry and who have completed Exploration 6.2 should be able to complete Exploration 6.3.

Try the sketch Cartesian (*Mac*) or Crtsian (*Windows*)—it's one of the many fun and useful supplementary sketches on the accompanying disk.

Exploration 6.1: Cartesian Coordinates

There are several ways of *graphing* (locating) points on a plane (a flat surface). You are probably familiar with graphing points in terms of *x*- and *y*-coordinates. The system of *x*- and *y*-coordinates is known as the *Cartesian coordinate system*, named after the man who invented it, the French mathematician and philosopher René Descartes. As the story is told, Descartes was sick in bed for several weeks and spent many hours watching a fly walk across the ceiling of his bedroom. He began wondering how he could describe the position of the fly as it moved across the ceiling.

In this exploration, you'll use Sketchpad to explore Descartes' solution to this problem.

Construct

For this exploration, you will need to construct a simple graph using Sketchpad. Construct an *x*- and a *y*-axis, and a point, then measure the *x*- and *y*-coordinates of the point.

- C1. Construct a horizontal line *x-axis*.

- C2. Construct a vertical line *y-axis*.

Point P needs to be in Quadrant I of this graph because Sketchpad does not measure negative distance easily.

- C3. Construct point *P* above line *x-axis* and to the right of line *y-axis*.

- C4. Measure the distances from point *P* to line *x-axis* and to line *y-axis*.

Investigate

- I1. Use the Selection Arrow tool to move point *P* around in the area above line *x-axis* and to the right of line *y-axis*. How does Sketchpad show you the *x*- and *y*-coordinates of point *P*?

- I2. Construct points *Q*, *R*, and *S* above line *x-axis* and to the right of line *y-axis*. Use point *P* to find the *x*- and *y*-coordinates of points *Q*, *R*, and *S*.

I3. Move points Q, R, and S so that they have the same x-coordinate but different y-coordinates. What do you notice?

I4. Move points Q, R, and S so that they have the same y-coordinate but different x-coordinates. Again, what do you notice?

 Conjecture

Record any insights you had during your investigation. What things that you believe are true in general are based on your specific observations? What patterns did you notice that you cannot explain yet? What questions would you like to explore further?

Follow-up—Exploration 6.1: Cartesian Coordinates

 Focus

F1. What will always be true of points with the same *x*-coordinate? What will always be the same of points with the same *y*-coordinate?

F2. Given any two perpendicular lines (lines *X-Axis* and *Y-Axis*) that meet at point *Origin*, write a precise definition of the *x*-coordinate and *y*-coordinate of any point in the plane.

F3. Does your definition allow for *negative* coordinates like (–3, 7) or (–10, –5)? If your definition does not account for negative coordinates, revise it.

 Look Back

L1. In the Focus section you made observations about points with the same *x*-coordinate and points with the same *y*-coordinate. Do your observations tell you anything about why graph paper looks the way it does?

L2. Based on your sketch and your knowledge of Cartesian graphing, what shape, do you think, was René Descartes' bedroom?

 Explore More

E1. How might your definition of Cartesian coordinates change if lines *X-Axis* and *Y-Axis* were not perpendicular. Use Sketchpad to explore your definition of *x*- and *y*-coordinates.

E2. What system of graphing might Descartes have invented if he slept in a circular room?

Teacher's Notes—Exploration 6.1: Cartesian Coordinates

Construct

C4. Students can measure the distance from a point to a line by constructing a perpendicular, or by selecting the point and the line and choosing Distance from the Measure menu.

Investigate

I2. It is sometimes difficult to move a particular object when it is placed on top of another object. This is especially true when one point is put on top of another. One simple solution is to use the Undo command on the Edit menu to move the points apart.

Conjecture

Notice that points with a common coordinate are collinear.

Focus

F2. This question provides a good opportunity to discuss the idea of precision in the wording of a definition.

Look Back

L2. This question works well with E2.

Explore More

E1. Interestingly, any two intersecting lines provide a workable coordinate system, although the way distance is defined needs to be changed in many cases.

Exploration 6.2: Polar Coordinates

It takes two numbers to describe the location of a point in a plane because a plane is two-dimensional. How many numbers does it take to describe a point in space?

The word *coordinate* literally means "equal in importance," as opposed to *subordinate*, which means "minor or secondary in importance." We talk about graphing with *x*- and *y*-coordinates because *x* and *y* are two numbers that have an equal importance in telling us the exact location of a point in a plane. The Cartesian coordinate system is one way to use two numbers of equal importance to give an exact location.

In this exploration you will explore another way to use two numbers to describe a point.

Construct

 Open the sketch Polar Coordinates (Mac) or Polar (Windows).

C1. Make sure that Sketchpad is measuring angles in directed degrees. Angle measure is set using the Preferences command in the Display menu.

○ P
Origin ○─────────
r coordinate = 2 cm
theta coordinate = 21°

C2. Construct points Q, S, T, and U anywhere in the sketch plane.

Investigate

I1. Move point P. How is Sketchpad computing the *r*-coordinate and theta-coordinate of point P?

Mathematicians often use Greek letters for angles. The symbol θ is the Greek letter theta (pronounced "thay'-tah"). It is easier to type out Theta than to make the symbol in Sketchpad.

You may find this a bit more difficult.

I2. Use point P to measure the *r*-coordinate and theta-coordinate of points Q, S, T, and U.

I3. Move points Q, S, T, and U so that they have the same theta-coordinate but different *r*-coordinates. What do you notice?

I4. Move points Q, S, T, and U so that they have the same *r*-coordinate but different theta-coordinates. Again, what do you notice?

Conjecture

Record any insights you had during your investigation. What things that you believe are true in general are based on your specific observations? What patterns did you notice that you cannot explain yet? What questions would you like to explore further?

Follow-up—Exploration 6.2: Polar Coordinates

Focus

F1. What will always be true of points with the same *r*-coordinate? What will always be true of points with the same theta-coordinate?

F2. In a polar coordinate system, we located points using a ray called the Theta Axis beginning at the origin and extending along the positive *x*-axis. Write a precise definition of the *r*-coordinate and theta-coordinate of any point relative to the Theta Axis in the polar coordinate system.

Look Back

L1. In the Focus section you made observations about points with the same *r*-coordinate and points with the same theta-coordinate. Use your observations to design a piece of polar graph paper.

L2. Based on your observations and your knowledge of Cartesian coordinates, do you think it is possible to have a *negative r*-coordinate in polar coordinates? Where would you locate a point with an *r*-coordinate of –2 and a theta-coordinate of 90° (1/2 radian)? If necessary, revise your definition of *r*- and theta-coordinates to account for negative values.

Explore More

E1. Can you create a piece of polar graph paper using Sketchpad? What would be appropriate spacing for lines of constant angle and lines of constant distance? Use Sketchpad to produce a piece of polar graph paper.

You might try looking in an atlas.

E2. Why do we use the word *polar* in polar coordinates? What uses might this system of coordinates have?

Teacher's Notes—Exploration 6.2: Polar Coordinates

Construct

C1. Choose Preferences from the Display menu to change the way Sketchpad measures angles.

Investigate

I2. It is sometimes difficult to move a particular object when it is placed on top of another object. This is especially true when one point is put on top of another. One simple solution is to use the Undo command on the Edit menu to move the points apart.

Conjecture

Notice that points with a common theta-coordinate lie on the same ray passing through the origin. Points with the same *r*-coordinate are all the same distance from the origin; they all lie on the same circle with center at the origin.

Focus

F2. The level of precision you expect probably depends on the level of your students. There are several possible definitions. One accurate definition that follows from the exploration is this: The point with coordinates (*r*, theta) is located at the intersection of a circle of radius *r* centered at the origin and the terminal ray of an angle that has the theta axis as its initial ray.

Look Back

L2. It is extremely difficult to explain the concept of negative *r*-coordinates succinctly. Rather than ask for a precise written definition (as in F2), you may want to stress students' ability to explain the concept verbally.

Explore More

E1. Polar graph paper made by students can be duplicated and used for other work on polar coordinates.

Exploration 6.3: Trigonometry and Coordinate Systems

Polar coordinates locate points by distance and angle, so it probably does not come as much of a surprise that people who study polar coordinates use trigonometry a great deal. Using trigonometry gives mathematicians and scientists a way to *translate* (convert) between polar and Cartesian coordinates. These two very different ways of locating points are linked by the basic trigonometric functions. Among other things, this makes it easier for scientists to study the orbits of planets and satellites.

Objects in the solar system move along paths defined by curves called *conic sections*, such as *ellipses* and *parabolas*. Conic sections are often studied using Cartesian coordinates, but a great deal of useful information about conic sections comes from studying their equations in polar coordinates.

In this exploration, you will investigate the relationship between polar and Cartesian coordinates, and you will discover a way to translate between the two coordinate systems.

Construct

For this exploration you will need to construct a sketch that shows the *x*-, *y*-, *r*-, and theta-coordinates of a movable point *P*.

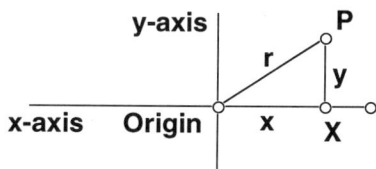

C1. Make sure that Sketchpad is measuring angles in directed degrees.

C2. Construct horizontal line *x-axis* and vertical line *y-axis*.

C3. Construct point *Origin* at the intersection of line *x-axis* and line *y-axis*.

C4. Construct point *P* above line *x-axis* and to the right of line *y-axis*.

C5. Construct a vertical line through point *P* perpendicular to line *x-axis*.

C6. Construct point *X* at the intersection of the vertical line through point *P* and line *x-axis*.

C7. Construct segment *r* from point *Origin* to point *P*.

C8. Construct segment *y* from point *P* to point *X*.

C9. Construct segment *x* from point *X* to point *Origin*.

Investigate

Polar coordinates are usually written in the form (r, theta), just as Cartesian coordinates are written (x,y).

I1. Use Sketchpad to determine the *x-, y-, r-,* and *theta-*coordinates of point *P*.

I2. Use your sketch to find the *Cartesian* coordinates of point *P* when its *polar* coordinates are (1, 30°).

I3. Find the Cartesian coordinates of point *P* when *r* is 1 and angle *theta* is the value of other *special angles* (0°, 45°, 60°, 90°, –30°, and so on). What do you notice?

I4. Move point *P* so that its Cartesian coordinates are (2, 3). Determine the polar coordinates (*r*, theta) for this location. Move point *P* so its polar coordinates are (2*r*, theta)—double the *r*-coordinate, but use the same theta-coordinate. What happens to the Cartesian coordinates?

I5. Repeat the last step of your investigation for other Cartesian coordinates. Try halving or tripling *r*.

 Conjecture

Record any insights you had during your investigation. What things that you believe are true in general are based on your specific observations? What patterns did you notice that you cannot explain yet? What questions would you like to explore further?

Follow-up—Exploration 6.3: Trigonometry and Coordinate Systems

Focus

F1. If you are given the polar coordinates of a point, how can you find the Cartesian coordinates without using Sketchpad? Write equations to find the x and y coordinates of a point given r and theta.

F2. If you are given the Cartesian coordinates of a point, how can you find the polar coordinates without using Sketchpad? Write equations to find the r- and theta-coordinates of a point given x and y.

Look Back

L1. Can you *prove* that the equations you wrote in the Focus section of this exploration are correct?

L2. Do the equations you wrote in the Focus section allow for negative values of x, y, r, and theta? If necessary, revise your equations.

Explore More

E1. Can you use your Cartesian–polar conversion equations to turn Cartesian equations into polar equations? Cartesian equations are usually written as $y = f(x)$, with y as a function of x. Polar equations are usually written as $r = f(theta)$, with r as a function of theta. What would the Cartesian equation for the line $y = 1$ look like as a polar equation? What about the equation $y = x^2$? Can you write down a method for turning Cartesian equations into polar equations?

When working with polar coordinates, theta is usually measured in radians rather than degrees. How would the Spiral of Archimedes change if you used degrees instead of radians?

E2. Can you use polar *coordinates* to graph polar *equations*? Polar equations are usually written as $r = f(theta)$, with r as a function of theta. That is, we calculate the value of r based on the value of theta. Try graphing the polar equation $r = theta$. This equation is called the Spiral of Archimedes. Do you see why? How will the picture change if you graph $r = 2theta$ instead? Compare this to the way $y = x$ and $y = 2x$ are similar or different.

Teacher's Notes—Exploration 6.3: Trigonometry and Coordinate Systems

Construct

C1. Choose Preferences from the Display menu.

Investigate

I3. Changing Preferences in the Display menu so that Sketchpad measures in inches may make it easier to complete this step.

I4. The wording of this question may be difficult for some students. Suggest they pick a Cartesian point such as (2, 3) and find its polar coordinates—in this case (3.6, 56.3°). Then double or triple r (make $r = 7.2$ or 10.8) and see what happens to the Cartesian coordinates. (They should change by the same factor.)

Conjecture

Students may notice that the sketch looks very much like the definition of the sine function given in terms of the coordinate system. Alternatively, they may use the right triangle to find values of sine and cosine.

Focus

There are several possibilities for the equations in this section involving sine, cosine, or tangent, and the Pythagorean Theorem.

Look Back

L2. You may want to encourage students to test their equations for negative values using graph paper and a calculator.

Explore More

E1. Polar equations often look "messier" to students than more familiar Cartesian equations. This provides an opportunity to discuss why different systems of coordinates are useful. One approach is to point out that certain kinds of equations are easier to graph in a given coordinate system. For example, circles are "messy" in Cartesian coordinates but very simple in polar coordinates.

E2. Students can use graphing calculators or graphing software to explore polar equations in more detail.